Tracking Animal Migration with Stable Isotopes

Volume 2 in the Academic Press | TERRESTRIAL ECOLOGY SERIES

Tracking Animal Migration with Stable Isotopes

Edited by

Keith A. Hobson and Leonard I. Wassenaar

Environment Canada
11 Innovation Blvd.
Saskatoon, Saskatchewan
Canada, S7N 3H5

AMSTERDAM • BOSTON • HEIDELBERG • LONDON
NEW YORK • OXFORD • PARIS • SAN DIEGO
SAN FRANCISCO • SINGAPORE • SYDNEY • TOKYO
Academic Press is an imprint of Elsevier

Front Cover credit: K. A. Hobson (insert photo) and K. A. Hobson and L. I. Wassenaar (background figure)

Academic Press is an imprint of Elsevier
84 Theobald's Road, London WC1X 8RR, UK
Radarweg 29, PO Box 211, 1000 AE Amsterdam, The Netherlands
Linacre House, Jordan Hill, Oxford OX2 8DP, UK
30 Corporate Drive, Suite 400, Burlington, MA 01803, USA
525 B Street, Suite 1900, San Diego, CA 92101-4495, USA

First edition 2008

ISBN: 978-0-12-373867-7
ISSN: 1936-7961

For information on all Academic Press publications
visit our website at elsevierdirect.com

Printed and bound by CPI Group (UK) Ltd, Croydon, CR0 4YY

Transferred to digital print 2012

Contents

Contributors

Numbers in parentheses indicate the pages on which the authors' contributions begin

STUART BEARHOP (129) Center for Ecology and Conservation, University of Exeter, Cornwall TR10 9EZ, United Kingdom

GABRIEL J. BOWEN (79, 129) Earth and Atmospheric Sciences Department and Purdue Climate Change Research Center, Purdue University, West Lafayette, Indiana 47907, USA, e-mail: gabe@purdue.edu

KEITH A. HOBSON (1, 45, 129) Environment Canada, Saskatoon, Saskatchewan S7N 3H5, Canada, e-mail: Keith.Hobson@EC.GC.CA

JEFFREY F. KELLY (129) Oklahoma Biological Survey and Department of Zoology, University of Oklahoma, Norman, Oklahoma 73019, USA, e-mail: jkelly@ou.edu

D. RYAN NORRIS (1, 107, 129) Department of Integrative Biology, University of Guelph, Guelph, Ontario N1G2W1, Canada, e-mail: rnorris@uoguelph.ca

LEONARD I. WASSENAAR (21, 129) Environment Canada, Saskatoon, Saskatchewan S7N 3H5, Canada, e-mail: Len.Wassenaar@EC.GC.CA

JASON B. WEST (79, 129) Department of Biology, University of Utah, Salt Lake City, Utah 84112-0840, USA, e-mail: jwest@biology.utah.edu

MICHAEL B. WUNDER (107, 129) Department of Fish, Wildlife, and Conservation Biology, Colorado State University, Fort Collins, Colorado 80523-1474, e-mail: mbw@lamar.colostate.edu

Acknowledgments

T his volume was made possible through the enthusiastic participation of the chapter authors, each of whom are specialists in very diverse fields. We are especially grateful to the numerous graduate students and colleagues that have, through their theses and publications and collaborations, indirectly or directly contributed to this consolidation of knowledge. We thank Professor Jim Ehleringer (University of Utah) for strongly encouraging this handbook on the topic of isotopes and migration. We thank Environment Canada for financial support. Phil Gregory designed the artwork for the cover. We especially thank Kristi Gomez and Kirsten Funk of Elsevier Academic Press for their assistance in the preparation of this volume.

Preface

The use of stable isotopes in the ecological and biological sciences is a rapidly growing area of research that was historically constrained within the domains of the earth and geological sciences. Today stable isotope measurements are becoming a routine, widespread, and integral component of many large-scale ecological and interdisciplinary environmental studies. Isotopic methods are invaluable for investigating the ecology of individuals and populations.

Here we focus on the very specific topic of using stable isotopes in unraveling terrestrial animal migration at various spatial scales around the globe. Migration is a compelling area of evolutionary and ecological research and understanding movement patterns in animals has become a topic of concern as we struggle to conserve threatened and other species that move across geopolitical boundaries. Previous traditional applications of tagging techniques have met with little success except for only a few larger and conspicuous species that can be intercepted with high probability. For the vast majority of migratory birds, mammals, and insects, more fundamental internal markers are required and, among these, stable isotopes show greatest promise if used appropriately.

The aim of this volume is to provide a college and graduate level handbook that will serve to introduce ecologists and biologists interested in terrestrial animal migration to the key elements of measuring, applying, and interpreting stable isotopes in terrestrial migratory systems. The reader will very quickly appreciate that this topic has not recently developed in isolation, but draws from many decades of foundational stable isotope research in the geological, hydrological, biological, and statistical sciences. A volume like this would not have been possible a few decades ago.

From an educational perspective, until recently there were no textbooks available for biologists and ecologists that provided a good overview of stable isotopes. Fortunately several textbooks have now appeared, and newcomers to the field of stable isotopes are also encouraged to consider reading books by B. Fry, *Stable Isotope Ecology* (2006); R. Michener and K. Lajtha, *Stable Isotopes in Ecology and Environmental Science* (2007), T. Dawson and R. Seigwolf, *Stable Isotopes as Indicators of Ecological Change* (2007) in addition to older benchmark books like P. W. Rundell, *Stable Isotope in Ecological Research* (1989). In addition, some internet websites are also of interest, primarily as an electronic forum for researchers using isotopes in migration. These include "Migrate" (www.migrate.ou.edu) and the biannual "Applications in Stable Isotope Ecology" international scientific meetings (www.isoecol.org).

Our hope is that the foundations laid here will provide a research stimulus for spawning new ideas and research into the fascinating topic and mysteries of animal migration in terrestrial and aquatic environments into the future.

Keith A. Hobson
Leonard I. Wassenaar

CHAPTER 1

Animal Migration: A Context for Using New Techniques and Approaches

Keith A. Hobson* and D. Ryan Norris[†]

*Environment Canada
[†]Department of Integrative Biology, University of Guelph

Contents

I. INTRODUCTION TO ANIMAL MIGRATION

The movement of organisms in space and time defines their interaction with their environments and, therefore, comprises a fundamental aspect of their ecology and evolutionary history. The degree to which organisms move also characterizes the range of resources they encounter, the array of hazards they experience from predators to hurricanes, and the degree to which they interact with other life forms. For animals, movement is very much a question of geospatial scale. While some species occupy a landscape of a few square meters for their entire lives or wander nomadically, others travel across thousands of kilometers in regular movements that constitute some of the most spectacular natural phenomena on the planet. These migrations are the movements that have captured the imagination of scientists and laypersons alike and leave us with a true sense of wonderment. Some key questions related to animal migratory movement include: How do these individuals know where they are going? How do they cope with the tremendous physical demands of travel? How do they adjust to unfamiliar or changing environments along the way to their destinations? How do they find their way back, often

Tracking Animal Migration with Stable Isotopes
K. A. Hobson and L. I. Wassenaar (Editors)
ISSN 1936-7961, DOI: 10.1016/S1936-7961(07)00001-2

to the precise location they left? Why do they move long distances in the first place? How is the abundance of populations influenced by where individuals go or where they are coming from?

Finding answers to these fundamental questions has proven to be an immense scientific challenge. A large part of this challenge has been the lack of tools available to scientists to infer or determine large- and small-scale animal movements. This volume explores recent developments in stable isotope methods that promise to contribute tremendously to the field of understanding animal movement in terrestrial ecosystems. However, before developing an understanding of those isotopic techniques, it is worthwhile exploring the scope of what we mean by migration.

The term *migration* often evokes images of spectacular seasonal movements of animals, especially birds, over vast distances. However, for thousands of years, where exactly birds went was one of the greatest natural mysteries of the world. In the *Iliad*, Homer believed that once cranes were done breeding in Greece, they flew to fight pigmies on the other end of the earth. In *On the History of Animals*, Aristotle famously hypothesized that [European] redstarts transformed themselves into Robins after the breeding season. Since redstarts leave Greece for Africa about the same time Robins arrive from their breeding sites further north, this seemed like a logical explanation. At the time, transmutation certainly seemed more plausible than birds traveling to other continents! Pliny the Elder (1855, 23–79 AD) was equally as inquisitive about seasonality in birds but a little more cautious. In Book X of his *Naturalis Historia*, he stated that "Up to the present time it has not been ascertained from what place the storks come, or whither they go when they leave us."

The sense of awe and mystery is no less relevant today as we are witness to some spectacular migration phenomena by a variety of organisms (Figure 1.1). The Arctic Tern (*Sterna paradisaea*) migrates from breeding grounds in the Arctic to wintering grounds in the Antarctic, an annual round trip of a staggering 40,000 km (Hatch 2002), equivalent to 2 years of day-to-day driving by the average North American motorist. Salmon return by the millions to natal streams at the end of their lives after spending years moving thousands of kilometers in the open ocean (Quinn 2005). In the first year of their lives, eastern populations of Monarch Butterflies (*Danaus plexippus*) in North America travel to overwintering roosts in the transvolcanic mountain range of central Mexico, a trip that can be over 4000 km for an insect that weighs only 500 mg. For those of us living in temperate environments, the annual spring and autumn movements of billions of migratory birds, from warblers to waterfowl, likely provide the most familiar examples of migration. All these migrations really involve individuals moving between two worlds (Greenberg and Marra 2005).

Although these movements may be spectacular, much confusion still exists as to what exactly we mean by migration. Clearly, migration can include both *to-and-fro* and *one-way* movements. A to-and-fro or *round-trip* migration can be characterized by animals either returning on the same path or by individuals following a *loop* migration pattern. Various other patterns of movement have been described between origin and destination, especially for birds. For example, *leap-frog* migration involves individuals at the northern limits of their breeding range traveling the farthest south (in northern hemispheric animals) to most distant wintering grounds whereas those from more southern breeding regions migrate the least distance to more northern wintering grounds (*e.g.*, Kelly *et al.* 2002). *Longitudinal* migration involves all individuals migrating the same approximate distance in a *chain* pattern or in *parallel* (Salomonsen 1955, Boulet and Norris 2006). Animal migration can be obligate whereby all members of a population move or facultative whereby resource availability may act to determine if migration occurs. *Partial migration* refers to a situation when only part of the population migrates. *Differential migration* describes those situations when migration patterns differ between sexes, age groups, or morphs within a population (Ketterson and Nolan 1976, Cristol *et al.* 1999).

One-way migration is probably a less common pattern of migration but is characteristic of many insects that move from the location they were produced to another location where they reproduce the next generation and die. That generation may then move on, repeating the process. In monarch butterflies, this occurs in a series of one-way steps involving multiple generations before a final cohort

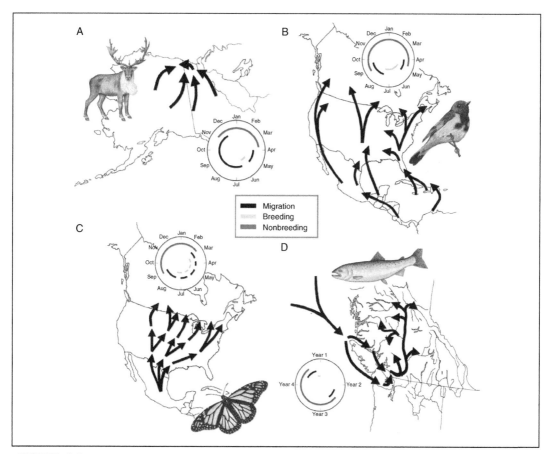

FIGURE 1.1 Examples of migration patterns in different taxa: (A) the Porcupine caribou herd (*Rangifer tarandus*), (B) songbirds (American redstart *Setophaga ruticilla*), (C) insects (Monarch butterfly *Danaus plexippus*), and (D) fish (Pacific salmon). Migratory pathways depicted by arrows are only generalizations.

returns to their starting point, overwinter roosts in the mountains of central Mexico. In other insects, such as the Black bean aphids (*Aphis fabae*), migration may involve a series of steps between host plants where individuals alternate between sexual and asexual reproduction (Cammell *et al.* 1989). Many salmon species also undertake spectacular migrations from the sea to freshwater rivers where they spawn and die. Although this is often considered one-way migration, it may be more accurately thought of as a "fatal round-trip migration" because offspring first have to migrate downriver where they spend several years at sea and then return to freshwater to spawn.

In mountainous habitats, animals may undergo *altitudinal* migrations where they seasonally track resource availability such as fruiting plants (Levey and Stiles 1992). These movements are more common in the tropics where frugivory is well developed and in more temperate areas of high relief. Although it may not seem like a form of migration, *nomadism*, where animals move in irregular patterns and breeding sites are established opportunistically where conditions are favorable, is a strategy employed by some animals. Several species of temperate forest seed-eating birds like crossbills, grosbeaks, and siskins are nomadic because of their reliance on cone crops or mast, which varies considerably in time and space (Newton 2006).

However, as Dingle (1996) noted, defining migration based solely on the outcomes (or patterns) such as to-and-fro, one-way, altitudinal, and nomadic movements can be somewhat problematic. For example, many animals also show to-and-fro or nomadic movements at a local scale within their *home range*. Regular movements within a home range can involve commuting between resource patches such as hummingbirds moving between sources of nectar or between resource patches and reproductive sites as seen in wolves during the denning period and in many species of colonial nesting seabirds. These types of movements are not considered migration partly because they do not involve a persistent movement between two areas. By persistent, we mean that individuals undertaking the act of migration are rarely distracted during the period of movement (Dingle 1996). Contrast this with foragers inside a home range who will typically take advantage of resources as they occur opportunistically. Migration also requires special physiological and behavioral adaptations during the period of movement, whereas movements within a home range usually require no such changes.

Using these patterns to define migration also leads to confusion because it encompasses another important type of one-way movement that is distinctive from migration: dispersal. Dispersal is the movement between the place of birth and a new breeding area (natal dispersal) or the movement between breeding sites (breeding dispersal; Greenwood and Harvey 1982). In the case of many migratory animals who move to other areas after breeding, dispersal may also be considered as the movement between nonbreeding sites (nonbreeding dispersal). Thus, in the most general sense, dispersal can be thought of as the one-way movement between sites within a predefined season or period of time. Unlike migration, the act of dispersal neither does necessarily involve a persistent movement to a predefined geographical target nor does necessarily require behavioral and physiological adaptations.

The line between dispersal and migration in migratory species can oftentimes be quite blurred. For example, consider an individual born at one location that migrates to a nonbreeding area and then migrates back to a different breeding site. In this case, the individual has both migrated (between breeding and nonbreeding sites) and dispersed (moved from one breeding site to another). Note that both migration and dispersal can be estimated without knowledge of the other process. For example, one could determine where an animal bred the previous year without knowledge of where it migrated to the previous season. Similarly, one may be able to determine where an individual migrated from the previous season without knowledge of where it bred the previous year. Knowledge of both processes is important for understanding the factors that influence individual fitness and population dynamics.

So, how do we define migration? A key underlying concept is that migration is typically in response to seasonal changes in resource abundance. As noted by Dingle and Drake (2007), migration can be considered an adaptation specific to arenas in which changes in habitat quality in different regions occur asynchronously so that movement allows a succession of temporary resources to be exploited as they arise. In this sense, migration involves escape and colonization. In their holistic view of migration, Dingle and coworkers see the arena as the environment to which migrants are adapted and the migration syndrome is the suite of traits enabling migratory activity that involves both locomotory capabilities and a set of responses to environmental cues that schedule and steer the locomotory activity. In this respect, migratory movements can be differentiated from movements within a home range or dispersal. Their conceptual model also includes a genetic complex that underlies the syndrome and a population trajectory comprising the route followed by the migrants, the timing of travel and stops along it, and the periods and locations corresponding to breeding and other key life-history events. The migration system involves a species' specific migration syndrome that determines how the animal responds to external cues so that preemptive movements can occur prior to resource collapse. A fascinating area of study involves understanding those physiological adaptations to migration, including the incorporation of reproduction and other tasks (Ramenofsky and Wingfield 2007) and the underlying genetic processes that result in selection for specific migratory traits (Pulido 2007, Roff and Fairbairn 2007).

As Dingle (1996) points out, the definition of migration must be based on the behavioral and physiological processes involved with the movement rather than the spatial pattern or outcome. Thus, we consider migration as the movement away from the home range that does not cease (at least not initially when suitable resources are encountered) which requires a set of behavioral and physiological adaptations for sustained movement that are unique from day-to-day adaptations related to self-maintenance and breeding.

II. MIGRATORY POPULATIONS, CONNECTIVITY, AND CONSERVATION

Although selection for adaptive traits related to migration is thought to occur at the individual level, migration systems and patterns are more often described at the population level, typically as the migration of populations or subpopulations within a species. We usually deal with population units when engaging conservation efforts and the conservation of populations that move over large distances presents considerable challenges (Webster *et al.* 2002, Marra *et al.* 2006). Populations of animals might be well protected and managed at one location, but then suffer no protection once they leave that area. For long-distance migrants, their travels may take them to new geopolitical jurisdictions where conservation measures are absent or inadequate. It makes little sense to expect migratory populations to respond positively to local conservation measures if key aspects of individual fitness or overall population health are being influenced at another location.

The problem of conserving species whose life-cycles cross geopolitical borders has become a profound issue in the twenty-first century as habitats everywhere are being reduced in size or quality, and migrants are forced to move among a myriad of patches, each with its own level of quality, safety, and prospects for future existence (Robinson *et al.* 1995, Moore *et al.* 2005). In this sense, migratory organisms are best considered to be made up of series of populations (that may or may not mix between different periods of the year) rather than as a single species per se. For some animals, populations breeding in one area may follow similar migration routes and winter in the same general wintering region. In this case, we would consider the subpopulations to show *strong connectivity*. Other breeding populations may disperse widely on the wintering grounds (or vice versa), which can be considered a case of *weak connectivity*.

The concept of migratory connectivity is important to conservation. Many forms of connectivity are possible (Webster and Marra 2005, Boulet and Norris 2006) ranging from the extreme cases where, for example, individuals of breeding populations show high philopatry to the same wintering location. Among birds, the Kirtland's Warbler (*Dendroica kirtlandii*) is a good example because the small breeding population in Michigan winters exclusively in the Bahamas. Weak connectivity may involve individuals from many breeding populations mixing among many wintering populations or vice versa. Quantifying patterns of connectivity can be difficult. One approach is to compare observed patterns of connectivity against the null hypothesis of complete mixing based on relative abundance (Norris *et al.* 2006). For example, let us say that a wintering population is sampled to estimate the geographic location of their breeding site. Suppose there are three potential breeding sites (A, B, and C) with the relative abundance of 0.6, 0.3, and 0.1, respectively. The null expectation for a sample of wintering birds would, therefore, be equal to the relative abundance of the breeding populations. Deviations from this expectation would indicate stronger than expected connectivity. The problem with this approach is that estimates of relative abundance are rarely available throughout a species entire range (all potential breeding sites).

Populations with strong connectivity are hypothesized to be those that are most vulnerable to declines because they have little chance to be "rescued" by peripherally connected populations (Marra *et al.* 2006). On the other hand, it is the strongly connected populations that may benefit the most from conservation efforts because the connections between two or more areas are well established. Although populations of weakly connected individuals may be inherently "safer," in the sense that risk is spread among more locations, these populations will understandably be harder to manage unless measures are enacted over large regions that encompass a significant percentage of the subpopulation of interest. It may also be more difficult to identify the causes of population declines in subpopulations that are weakly connected. For example, when migratory populations are mixed between two periods, habitat loss in one subpopulation may produce synchronous declines in subpopulations in the following season (Marra *et al.* 2006).

Knowledge of migratory connectivity is also important for being able to predict how conservation measures in one season will influence populations the following season. For example, Martin *et al.* (2007) demonstrated that the protection of nonbreeding habitat for American redstarts (*Setophaga ruticilla*) based on rate of habitat loss, density, and land cost resulted in the almost complete extinction of one breeding region. This was because the criteria for deciding where and when to conserve nonbreeding habitat did not consider where individuals were spending the remainder of the annual cycle. Consideration of how populations are connected can, therefore, result in radically different decisions on how to allocate resources for conserving species.

In many ways, most migratory animals have evolved to maintain ambient conditions as constant as possible throughout the annual cycle. This contrasts with resident species in temperate areas that must adjust seasonally to extreme changes in weather and foraging opportunities. As such, it is debatable whether or not migrants are inherently more adaptable or plastic in their behaviors compared to nonmigrants. This issue is particularly interesting in view of global anthropogenic and natural environmental, landscape, and climatic changes that have occurred over decadal and millennium scales and will continue into the future (Crick 2004, Møller *et al.* 2004). For migratory birds, it is not clear whether they will be able to quickly adapt to anthropogenic changes wrought on a decadal timescale (Møller *et al.* 2004).

Timing of breeding has important implications for avian productivity and is one of the traits most noticeably affected by spring weather conditions (Dunn and Winkler 1999, Winkler *et al.* 2002, Dunn 2004). Birds that nest earlier generally produce more offspring that are ultimately recruited into breeding populations (Perrins 1970, Dunn 2004). It remains uncertain whether migratory birds and other animals will be able to continue to synchronize the timing of their breeding with optimal conditions for reproduction on breeding areas in light of changes in climate and land use (Visser *et al.* 1998, 2004, Both and Visser 2001, Sanz *et al.* 2003).

In their long-distance travels that cross international borders, migratory animals can also act as vectors of parasites and pathogens. There are a number of ornithophylic mosquitoes that can infect migratory birds that in turn act as introductory or amplifying hosts (Rappole *et al.* 2000, Reed *et al.* 2003). Recent interest in the spread of avian influenza viruses around the globe has increased the interest in determining the migratory origins of infected and noninfected wild birds (Liu *et al.* 2005). This level of interest will only increase as the global climate and the distributions of suitable habitats available to these pathogens change over time.

Effective management of species also requires identifying where populations are most productive. We are all familiar with field guides that show the absolute spatial ranges occupied by a given species, but where within that vast space are the most young being produced and recruited into the autumn migratory population and where are individuals experiencing the highest survival rates during the nonbreeding period? Understanding where hot spots of productivity and survivorship occur helps us protect those places and better understand factors that may be limiting populations. Nor are species necessarily more productive where they are most abundant, and numerous examples exist of how animals can be lured into ecological traps where they suffer reduced success. For migratory game birds, understanding how zones of productivity relate to protected areas and the timing of harvest is particularly crucial to their conservation.

For many years, our understanding of the ecology of migrant organisms was understandably influenced by their activities during the breeding season in locations where they were accessible to scientists working in the temperate regions. Such interest was also driven by an assumption that decisions made by individual animals related to breeding were of paramount importance to an individual's fitness. Aspects of breeding success related to habitat choice, mate choice, timing of breeding, and the quality of individuals were topics that were considered largely as they played out on the breeding grounds, with little regard to events that preceded and followed them on the wintering grounds and during migration. This situation changed in the 1980s when waterfowl researchers began to discover that female body condition upon arrival on the breeding grounds in spring was linked to conditions she experienced in winter, and that these could have profound influence on the number of young recruited into the autumn population (Heitmeyer and Fredrickson 1981, Kaminski and Gleusing 1987).

In the late 1990s, it has become more widely accepted that events occurring outside the breeding area can profoundly affect the fitness of migratory individuals. Much of this recent interest has been generated from the work of Marra *et al.* (1998) who, using stable isotopes, first demonstrated in American redstarts a clear link between spring arrival times of males and the quality of the habitat they occupied in winter. The quality of a habitat where an animal winters will determine the level and speed with which the necessary body condition for migration can be reached, and so can determine when and in what condition that individual arrives on the breeding grounds. Several studies of birds have shown that earlier arriving birds can obtain better territories, initiate clutches sooner, and generally fledge more young than those arriving later (*e.g.*, Gill *et al.* 2001, Bêty *et al.* 2003, Norris *et al.* 2004). How costs associated with reproduction might influence an individual's ability to compete during the nonbreeding season remains a mystery. Nonetheless, seasonal interactions between the wintering grounds and the breeding grounds, as well as during migration itself, should ultimately play a large role in determining lifetime fitness and population abundance (Norris and Marra 2007).

Establishing details of migratory programs in animals is fundamental to understanding key aspects of their evolution, life history, and conservation. The field has witnessed a renaissance in recent years primarily because the development of new analytical techniques has provided some true breakthroughs. There is also a more urgent sense that we need to more quickly understand key migratory linkages for animal populations in order to help conserve them in a rapidly changing world. It is in this spirit that we developed this volume on stable isotope techniques because that approach has and will continue to contribute immensely to the study of animal migration. We also recognize that, as with all techniques and approaches, there is a real need to provide a firm foundation for practitioners and to point to ways in which the field might best progress.

III. SCIENTIFIC TOOLS USED TO STUDY MIGRATION

A. Extrinsic Markers

Tracking migratory terrestrial animals has involved numerous techniques over the years (Table 1.1). Until very recently, all approaches involved the use of extrinsic markers applied to individuals with the hope of relocating those same individual elsewhere, or on the use of recognized phenotypic or morphological traits that showed known geographic variation. For birds, there is a rich literature on geographic variation in plumage and morphology and these traits have been used to describe migratory connectivity. A good example is the Yellow-rumped Warbler (*Dendroica coronata*) that shows plumage variation between eastern (*coronata*) and western (*audubon*) races in North America and that also segregates largely on the wintering grounds. Similarly, the Swainson's thrush (*Catharus ustulatus*) occurs as a reddish backed phenotype on the Pacific coast of North America and a more common olive-backed from elsewhere. Birds captured during migration can readily be placed into these types without the need for mark-recapture techniques. Most current field guides provide numerous examples of geographic variation in plumages (Sibley 2001). An excellent example of how plumage and morphometrics have been used to describe migration patterns in Nearctic–Neotropical migrant birds that moved through Mexico was that of Ramos and Warner (1980) and Ramos (1983). However, typically the geographic resolution provided by this approach is both poor and highly variable among taxonomic groups. For example, as noted by Boulet and Norris (2006), within the 50 species of migratory wood warblers of the New World, 65% are monotypic, 30% include 2–4 subspecies, and only 6% include 5 or more subspecies.

TABLE 1.1 Summary of various techniques used to track migration in animals

Technique	Advantages	Disadvantages	References
Extrinsic	1. can apply to a broad range of animals, 2. high spatial resolution.	1. requires initial capture and then recapture, 2. biased toward initial capture population	
Phenotypic variation (*e.g.*, morphology, plumage)	1. inexpensive, 2. can be applied to historical specimens with high degree of confidence.	1. low resolution, 2. not applicable to all species, 3. provides estimate of natal origin only	Ramos (1983), Ramos and Warner (1980), Bell (1997), Boulet and Norris (2006)
Banding/marking	1. inexpensive, 2. provides exact information on start and end of movements.	1. many species have low recovery rates (<0.5%), 2. can take many years to get adequate data, 3. still only a small number of major banding stations across globe, 4. both marking and recovery (start and end points) are biased toward locations of major banding stations.	Brewer *et al.* (2000), Bairlein (2001), Berthold (2001)
Radio transmitters	1. produces precise locations (relative to extrinsic markers through triangulation, 2. can obtain precise trajectory if within range of transmission.	1. low range 2. expensive relative to banding or morphologial methods, 3. Some evidence suggests transmitters can have adverse effect on behavior	Vega-Rivera *et al.* (1998), Cochran *et al.* (2004)
Satellite transmitters	1. precise animal trajectory.	1. expensive (severely constrains sample size), 2. constrained to large animals only (~300 g), 3. some evidence suggests that transmitters can have adverse effect on behavior.	Britten *et al.* (1999), Hays *et al.* (2001), Berthold *et al.* (2002)
ICARUS Project	1. small and light transmitters allow many types of species to be tracked, similar to statellite tages, can track individuals over large distances (global),	1. huge start-up investment, 2. not yet proven technology,	Wikelski *et al.* (2007)

Method	Advantages	Disadvantages	References
	3. inexpensive following start-up investment.	3. like above, some evidence suggests that transmitters can have adverse effect on behavior.	
passive radar	1. coverage over large geographic area, 2. inexpensive, 3. individuals do not have to be captured.	1. coverage only from existing stations or portable instruments, 2. poor ability to determine species- and individual-specific movements	Bruderer (1997), Gauthreaux and Belser (2003)
Transponders	1. small transponder size.	1. Requires external (radar/microwave) activation, 2. low range, 3. coverage only from existing stations or portable instruments.	Riley et al. (1996)
Archival geolocation tags	1. produces an animal trajectory; interval between recorded locations can be customized, 2. light weight (as low as 1.5 g), 3. inexpensive relative to satellite tags (but more expensive than radio transmitters)	1. individuals must be recaptured to download data, 2. accuracy relative to satellite tags still low (~200 km)	Shaffer et al. (2005, 2006)
Intrinsic	because only requires one capture, it is: 1. not biased to initial capture population, 2. less labor intensive than most extrinsic methods.	1. biased to final captured population (can be overcome by comprehensive sampling coverage), 2. typically lower resolution than extrinsic markers.	Hobson (1999), Webster et al. (2002), Rubenstein and Hobson (2004)
Contaminants	1. potentially high spatial resolution (e.g., Mirex).	1. lack of a priori maps of distribution and relative abundance, 2. distribution of containments may be vary predictably over geographic areas, 3. potential transport of contaminants may dampen or provide unreliable geographic signal.	Ochoas-Acuna et al. (2002), Braune and Simon (2003)
Parasites Genetics	1. several markers possible,	1. species specific,	Fallon et al. 2006 Kelly et al. (2005), Smith et al. (2005), Boulet et al. (2006)

(continued)

Table 1.1 *(continued)*

Technique	Advantages	Disadvantages	References
	2. *east vs west* resolution of migratory fauna in North America.	2. typically low resolution,	
Trace elements	1. simultaneous measurement of a large number of elements,	3. provides estimate of natal origin only.	Parrish *et al.* (1983), Kelsall (1984), Szep *et al.* (2003), Norris *et al.* (2007)
	2. potentially high spatial resolution.	1. lack of a priori maps of distribution and relative abundance,	
		2. expensive,	
		3. requires more sample tissue than either isotopes or genetic markers,	
		4. requires sampling of tissue that is metabolically inert after growth,	
		5. spatial resolution could be too high (*i.e.*, requires sampling of all possible populations),	
		6. evidence that some elements may be integrated into metabolically inactive tissues after growth is complete.	
Stable isotopes	1. inexpensive,	1. often low resolution,	Chamberlain *et al.* (1997), Hobson and Wassenaar (1997), Hobson (1999, 2005), Norris *et al.* (2006), this volume
	2. not species or taxon specific,	2. lack of isotopic basemaps for several elements,	
	3. multiple isotopes can be combined to increase spatial resolution.	3. ideal tissue is metabolically inactive after growth,	
		4. for metabolically active tissue, turnover rate of elements unknown for many tissues/species,	
		5. Animal physiology may influence dD values of body water and so complicate interpretations.	

By far the most widespread approach to tracking migrant animals is through the application of passive extrinsic markers. For birds, these have overwhelmingly involved leg bands or rings carrying a unique number combination and some instruction on where to report the band if it is recovered. Other markers such as patagial tags, numbered neck collars, streamers, or color dyes have also been used. Insects have proven to be more of a challenge due to their small size, but numbered labels have been successfully affixed to the wings of butterflies and later recovered (www.monarchwatch.org). In the relatively short historical period (*i.e.*, 100 year) of banding birds, many millions of individuals have been individually tagged. For a number of species with small global populations and restricted ranges, some very impressive recovery rates have been achieved (*e.g.*, Owen and Black 1989) and some key insights into migratory connectivity established (Gill *et al.* 2001). However, for the vast majority of species, extremely low recovery rates (*i.e.*, $<0.01\%$) are the norm (Hobson 2003).

B. Transmitters, Radar, and Satellites

Active extrinsic markers are those that send out signals that can be intercepted with a suitable receiver device. Advances in transmitter technology have allowed the placement of devices on migratory animals. If a receiver is within range of the transmitter, then the location of the animal can be inferred either by tracking down the individual or by triangulation with more than one receiver. Radio-frequency transmitters can be made small enough (0.5 g or lower, see below) to place on small passerines and bats. However, with miniaturization comes a loss of both range and battery life so that optimally, these devices provide location information up to a few kilometers. Nonetheless, adventurous researchers have attempted to follow migrating birds, bats, and even dragonflies equipped with transmitters over at least portions of their flight paths (Wikelski *et al.* 2003, Cochran *et al.* 2004, Holland *et al.* 2006). Using a combination of aerial and ground tracking receivers, Wikelski *et al.* (2006) followed the southward migration of dragonflies (Common green darner *Anax junius*) equipped with 300 mg radio transmitters. Presumably, placement of automatic receivers, such as one that is already in place on Barro Colorado Island, Republic of Panama (M. Wikelski, personal communication), could monitor the presence of an individual at a few key locations like stopovers or areas where migrants are concentrated, but such an approach is still a bit like trying to locate a needle in a haystack. Cell phone technology has also provided new possibilities for tracking animals but the miniaturization and development required for animal tracking is a hurdle. As well, cell phone network coverage is limited at a global scale and locational accuracy is low (Stokely 2005). A final concern is that the fastening of markers or transmitters may alter flight or movement behavior.

Radar technology has made great contributions to our understanding of migration because this provides information of flying animal movements over considerable distance (Gauthreaux and Belser 2003). However, like automated receivers, radar installations are often fixed, and mobile radar systems are generally impractical to follow animal movements over migratory distances. Crossband transponders placed on migratory organisms can be used to elicit a detectable radio frequency signal after the transponder is intercepted by radar. The transponder is dormant until such an event and so can last for a much longer period than conventional "always transmitting" radio frequency tags. However, such *active* radar systems face a number of limitations, again related to size and weight of instrumentation and the need to intercept the organism of interest within range of the radar system (see www.earthspan.org).

Satellite transmitters have provided a major advance in methods to track migratory animals because they provide extremely accurate positions of individuals remotely. Precise trajectories of individuals are possible over vast portions of the globe that have suitable satellite coverage. Satellite tracking systems such as ARGOS (www.argosinc.com) have been placed on different satellites from the US National Oceanic and Atmospheric Administration, The Japanese Space Agency, and the European Meteorological Satellite Organization. The ARGOS system collects data from platform terminal transmitters and delivers telemetry data back to the user. Unlike other extrinsic techniques, this approach does not require the physical capture of individuals once they are marked (although many researchers attempt this in order to recover the expensive transmitters). One of the most spectacular examples of this approach was provided by Jouventin and Weimerskirch (1990) who tracked Wandering albatrosses (*Diomedea exulans*) on foraging flights up to 15,000 km. Future prospects for this technology are encouraging and there is strong interest in diminishing the size of transmitters and batteries so that smaller species can be monitored. Currently, the smallest satellite tags available are about 9.5 g and potentially could be used on 240 g animals (using the <5% body weight rule, Murray and Fuller 2000). Unfortunately, this still excludes about 81% of the world's birds and 68% of the mammals and, of course, all migratory insects (Wikelski *et al.* 2007).

The use of satellites may potentially provide a major breakthrough in tracking migratory animals down to the size of large insects. Wikelski *et al.* (2007) have proposed that a new satellite equipped with radio receivers could track radio tags with radiated power as low as 1 mW with an accuracy of a few kilometers under favorable conditions. This power can be achieved from existing radio frequency tags as small as <1 g. This approach is similar to solutions derived by space researchers who need to detect very weak signals from distant galaxies and so the technology exists to modify the solution to detecting weak radio signals from earth amidst a background of much stronger radio frequency noise. A group known as International Cooperation for Animal Research Using Space (ICARUS, www.icarusinitiative. org/solutions) is promoting this idea as a means of tracking small animals around the globe. They estimate that it will take about US$50–100 million in order to build and launch a satellite.

A relative new application for tracking migratory animals is the use of archival geolocation tags (Shaffer *et al.* 2005, 2006). These tags estimate longitude and latitude based on light levels and sea-surface temperatures, and long-distance movement data have now been reported on shearwaters (Shaffer *et al.* 2005) and albatross (Shaffer *et al.* 2006) with several more species currently being tested. The advantage of these tags is that they are significantly lighter than satellite transmitters (now 1.5–3 g). However, the disadvantage is that their accuracy, using both combined light levels and sea-surface temperatures, is only ±200 km that restricts application of these tags to questions pertaining only to large-scale movements.

With the exception of satellite transmitters, all extrinsic markers require that the individual be recaptured, resighted, or at least move within detection distance at a later time. The probability of this recapture will clearly be the product of a number of individual probabilities related to the number of observers, the likelihood of an observer reporting the information, the regions and habitats used by the animal, and so on. As we have seen, combining these probabilities can result in vanishingly small chances of obtaining information on any given individual.

Apart from the burden of recovering a marked individual, the use of passive extrinsic markers suffers another fundamental flaw—they provide information only on the movement of marked individuals. The possibility of extrapolating the findings based on a small marked cohort to the population or species level depends on how representative the marked individuals are. Our confidence in this approach increases with the number of independent recaptures and some band recovery data show unequivocal patterns that are clearly representative of populations. A single band recovery or satellite track while interesting and ultimately useful may tell us very little about what populations are doing.

IV. INTRINSIC MARKERS

The primary advantage of intrinsic markers is that initial marking of individuals is not necessary and that every capture provides information on origin. In this sense, every capture becomes a recapture. The sampling scheme is biased then, only by the limitations of where animals are ultimately located and this typically represents a much less serious form of bias compared to where individuals can be marked initially using extrinsic markers.

A. Contaminants, Parasites, and Pathogens

There are several forms of potential intrinsic markers that can be used to infer origins of migratory animals. If the spatial distribution of a suite of contaminants were known, then presumably the occurrence of contaminants in animals could provide some information of the past distribution of the animals being sampled. Although this approach has not yet been used to any significant degree, there is some interesting potential here. For example, Braune and Simon (2003) noted that in contrast to Thick-billed Murres (*Uria lomvia*) and Black-legged Kittiwakes (*Rissa tridactyla*), the pattern and concentration of dioxin/furan congeners as well as the ratio of total PBDEs to HBCD (hexabromocy-clododecane) in Northern Fulmars (*Fulmarus glacialis*) breeding in the Canadian high Arctic was suggestive of them wintering in the northeast Atlantic. In the Great Lakes region of North America, mirex could be used as a marker to distinguish among those waterbirds breeding and wintering in different regions because the upper lakes have much lower concentrations of this contaminant than the lower lakes. The ratio of DDE to PCB can also be used to illustrate which birds have likely overwintered in agricultural areas of South/Central America versus coastal areas (B. Braune, unpublished data). Brominated flame retardants are another material that occurs in greater concentrations in food webs used by animals exposed to European air masses than those exposed to North American air masses (de Wit 2002). Finally, methyl mercury and other heavy metals are another potential marker in migratory animals because exposure to these varies considerably throughout the world (Hario and Uuksulainen 1993, Janssens *et al.* 2002, Ochoas-Acuna *et al.* 2002).

Similar to contaminants, parasite and pathogen exposure experienced by migratory animals varies geographically and there is interest in using these markers to examine movements of assayed individuals (Ricklefs *et al.* 2005). Very little research has been conducted on these sorts of tools to assist with deciphering animal movements, perhaps because we have little information to begin with on the spatial distributions of the potential markers themselves. On the other hand, the use of genetics has generated more interest because it is possible to assay and describe genetic variation across breeding populations and then to use this information to assign a probability that a given individual came from a given (known) subpopulation (Smith *et al.* 2005). The identification of population structure using genetic markers has included the use of allozymes, mitochondrial DNA sequences, and DNA fragment analyses such as microsatellites and amplified fragment length polymorphism. These markers can yield different scales of population structure because they evolve at different rates. In North America, genetic markers have been particularly useful in differentiating between eastern and western breeding origins of wintering Neotropical migrant songbirds (Smith *et al.* 2005, Boulet *et al.* 2006), reflecting patterns of rapid demographic expansions following glaciation events on that continent. The combination of marker techniques that can provide information on north–south origins in North America would clearly augment the resolution of genetic approaches. Genetic analyses also hold great potential for developing parasite markers in migrant organisms because PCR-based assays of blood parasites can now identify pathogens to species and haplotype (Ricklefs *et al.* 2005).

B. Trace Elements

Trace elements are another set of intrinsic markers that can potentially be used to track individual movements over both small and large geographic distances. Similar to stable isotopes, the idea with trace elements is that individuals acquire distinctive chemical profiles at one geographic location and then carry that profile with them to another area where they can be sampled to estimate their geographic origin. In the past, one of the limitations of using trace elements was the amount of sample required for analysis. However, the use of inductively coupled plasma mass spectrometers has allowed smaller quantities of samples (3 mg: T. K. Kyser, personal communication; Donavan *et al.* 2006, Norris *et al.* 2007) to be measured with relatively high precision. This technological advance has allowed researchers to focus on nondestructive tissues, such as metabolically inactive keratin, that are grown during specific periods of the migratory cycle. Trace element profiles have been used to infer whether individuals sampled in the same breeding population originate from different places the previous winter (Szep *et al.* 2003).

However, because we have little idea of how trace elements vary over the landscape, the ability of this technique to estimate the precise origin of individuals seems, at the moment, limited. To determine how trace elements vary across the landscape, Norris *et al.* (2007) measured 42 trace elements in feathers of western sandpipers (*Calidris mauri*) at 5 different locations on their tropical wintering grounds. Feathers were grown during the winter periods and were, therefore, assumed to provide a signature of known-origin. Elemental profiles successfully distinguished between birds wintering at all five locations. However, two locations were less than 3 km apart suggesting that trace elements are likely very specific to the location or origin in which they were sampled. Trace element approaches are probably best suited to species that are aggregated over only a few breeding or wintering sites so the majority of populations can be sampled over the entire range (Donavan *et al.* 2006). Studies several years ago also show that some trace elements in feathers may be acquired after growth (Bortolotti and Barlow 1988, Bortolotti *et al.* 1988), implying that this may not be a reliable approach for tracking long-distance movements. Further studies are needed to test the generality of these results before trace elements can be widely used as a marker of geographic location.

C. Stable Isotope Approaches

The intent of this volume is to provide the reader with a comprehensive background needed to understand the application of stable isotope tools to the study of animal migration. This is a new and exciting field and we are seeing a rapid increase in the number of researchers using this approach. However, animal ecologists have really only recently become aware of using stable isotope measurements in ecological studies in general. The world of stable isotopes is a multidisciplinary one that has a very rich history involving physics, earth sciences, biogeochemistry, animal and plant physiology, and even anthropology and archaeology. The 1988 publication of the book *Stable Isotopes in Ecological Research* edited by Rundel *et al.* (1988) marked a significant pivot point that informed ecologists of the immense potential of stable isotope measurements in natural ecosystems. This was quickly followed up with *Stable Isotopes in Ecology and Environmental Science* edited by Lajtha and Michener (1994), which is now updated to Michener and Lajtha (2007). Most recently, Fry (2006) has produced a useful text that provides some of the necessary background to understanding fundamental concepts in stable isotope ecology and Karasov and Martínez del Rio (2007) provide a very useful overview of key linkages between animal physiology and isotopic ecology.

In Chapter 2, the essential background required to understand the behavior of stable isotopes in nature, their measurement, and the fundamentals of mass spectrometry is provided. This chapter will provide the reader with a good starting point for stable isotope applications as well as the field of tracking migration. Chapter 3 introduces fundamental concepts and principles underlying isotopic tracking of migration and provides a perspective on the history of this approach in terrestrial systems. Key examples are used to illustrate specific concepts but the reader is also provided with a detailed summary of most work to date. Many examples will include birds. In part, this reflects the fact that avian biologists have been quick to apply isotopic techniques to a number of facets of bird ecology. Birds are also well studied relative to other taxa and provide some of the best subjects for the study of migration in vertebrates.

In Chapter 4, isotopic patterns across landscapes are reviewed that can be used to infer animal origins. These "isoscapes" provide the fundamental templates upon which migration is played out. This is a current area of exciting research involving remote sensing and Geographic Information Systems to portray expected spatial patterns. In Chapter 5, the numerous ways in which isotopic data can be analyzed to infer the origin of individuals are reviewed and analyzed. Because methods of assigning individuals have vastly improved over the last few years, suggestions for researchers embarking on designing studies to track migratory animals using stable isotopes are also provided. Finally, Chapter 6 combines the thoughts of all authors and others to help establish the way forward.

V. REFERENCES

Bairlein, F. 2001. Results of bird ringing in the study of migration routes. *Ardea* **89**:7–19.

Bell, C. P. 1997. Leap-frog migration in the Fox Sparrow: Minimizing the cost of spring migration. *Condor* **99**:470–477.

Berthold, P. 2001. *Bird Migration. A General Survey*. Oxford University Press, Oxford.

Berthold, P., W. V. D. Bosch, Z. Zakubiec, C. Kaatz, M. Kaatz, and U. Querner. 2002. Long-term satellite tracking sheds light upon variable migration strategies of White Storks (*Ciconia ciconia*). *Journal of Ornithology* **143**:498–493.

Bêty, J., G. Gauthier, and J.-F. Giroux. 2003. Body condition, migration and timing of reproduction in Snow Geese: A test of the condition-dependent model of optimal clutch size. *American Naturalist* **162**:110–121.

Both, C., and M. E. Visser. 2001. Adjustment to climate change is constrained by arrival date in a long-distance migrant bird. *Nature* **411**:296–298.

Boulet, M., and D. R. Norris. 2006. The past and present of migratory connectivity. *Ornithological Monographs* **61**:1–13.

Boulet, M., H. L. Gibbs, and K. A. Hobson. 2006. Integrated analysis of genetic, stable isotope, and banding data reveal migratory connectivity and flyways in the northern Yellow Warbler (*Dendroica petechia*; *Aestiva* group). *Ornithological Monographs* **61**:29–78.

Bortolotti, G. R., and J. C. Barlow. 1988. Stability of mineral profiles of spruce grouse feathers. *Canadian Journal of Zoology* **66**:1948–1951.

Bortolotti, G. R., K. J. Szuba, B. J. Naylor, and J. F. Bendell. 1988. Stability of mineral profiles of spruce grouse feathers. *Journal of Wildlife Management* **52**:736–743.

Braune, B., and M. Simon. 2003. Dioxins, Furans, and non-ortho PCBs in Canadian Arctic seabirds. *Environmental Science and Technology* **37**:3071–3077.

Brewer, A. D., A. W. Diamond, E. J. Woodsworth, B. T. Collins, and E. H. Dunn. 2000. *The Atlas of Canadian Bird Banding, 1921–95. Volume 1: Doves, Cuckoos and Hummingbirds Through Passerines*. Canadian Wildlife Service Special Publication, Ottawa, Ontario, CanadaURL:http://www.cws-scf.ec.gc.ca/publications/spec/atlas_e.cfm.

Britten, M. W., P. L. Kennedy, and S. Ambrose. 1999. Performance and accuracy evaluation of small satellite transmitters. *Journal of Wildlife Management* **63**:1349–1358.

Bruderer, B. 1997. The study of bird migration by radar. *Naturewissenschaften* **84**:45–54.

Cammell, M. E., G. M. Tatchell, and I. P. Woiwod. 1989. Spatial pattern of abundance of the Black bean aphid, *Aphis fabae*, in Britain. *Journal of Applied Ecology* **26**:463–472.

Chamberlain, C. P., J. D. Blum, R. T. Holmes, X. Feng, T. W. Sherry, and G. R. Graves. 1997. The use of stable isotope tracers for identifying populations of migratory birds. *Oecologia* **109**:32–141.

Cochran, W. W., H. Mouritsen, and M. Wikelski. 2004. Migrating songbirds recalibrate their magnetic compass daily from twilight cues. *Science* **304**:405–408.

Crick, H. Q. P. 2004. The impact of climate change on birds. *Ibis* **146**:48–56.

Cristol, D. A., M. B. Baker, and C. Carbone. 1999. Differential migration revisited: Latitudinal segregation by age and sex class. *Current Ornithology* **15**:33–88.

de Witt, C. 2002. An overview of brominated flame retardants in the environment. *Chemosphere* **46**:583–624.

Dingle, H. 1996. *Migration: The Biology of Life on the Move.* Oxford University Press, New York.

Dingle, H., and V. A. Drake. 2007. What is migration?. *Bioscience* **57**:113–121.

Donavan, T., J. Buzas, P. Jones, and H. L. Gibbs. 2006. Tracking dispersal in birds: Assessing the potential of elemental markers. *Auk* **123**:500–511.

Dunn, P. O. 2004. Breeding dates and reproductive performance. *Advances in Ecological Research* **35**:69–87.

Dunn, P. O., and D. W. Winkler. 1999. Climate change has affected the breeding date of tree swallows throughout North America. *Proceedings of the Royal Society of London, Series B, Biological Sciences* **266**:2487–2490.

Fallon, S. M., R. C. Fleisher, and G. R. Graves. 2006. Malarial parasites as geographical markers in migratory birds?. *Biology Letters* **2**:213–216.

Fry, B. 2006. *Stable Isotope Ecology.* Springer, New York.

Gauthreaux, S. A., and C. G. Belser. 2003. Radar ornithology and biological conservation. *Auk* **120**:266–277.

Gill, J. A., K. Norris, P. M. Potts, T. G. Gunnarsson, P. W. Atkinson, and W. J. Sutherland. 2001. The buffer effect and large-scale population regulation in migratory birds. *Science* **412**:436–438.

Greenberg, R., and P. P. Marra (Eds.). 2005. *Birds of Two Worlds: The Ecology and Evolution of Temperate-Tropical Migration Systems.* Johns Hopkins University Press, Baltimore, MD.

Greenwood, P. J., and P. H. Harvey. 1982. The natal and breeding dispersal of birds. *Annual Review of Ecology and Systematics* **13**:1–21.

Hario, M., and J. Uuksulainen. 1993. Mercury load according to moulting area in primaries of the nominate race of the Lesser Black-backed Gull. *Larus f. fuscus. Ornis Fennica* **70**:32–39.

Hatch, J. J. 2002. Arctic Tern (*Sterna paradisaea*). *In* A. Poole and F. Gill (Eds.) *The Birds of North America,* **No. 707.** The Birds of North America, Inc., Philadelphia, PA.

Hays, G. C., S. Akesson, B. J. Godley, P. Luschi, and P. Santidrian. 2001. The implications of location accuracy for the interpretation of satellite-tracking data. *Animal Behaviour* **61**:1035–1040.

Heitmeyer, M. E., and L. H. Fredrickson. 1981. Do wetland conditions in the Mississippi Delta hardwoods influence mallard recruitment? *Transactions of the North American Wildlife and Natural Resources Conference* **46**:44–57.

Hobson, K. A. 1999. Tracing origins and migration of wildlife using stable isotopes: A review. *Oecologia* **120**:314–326.

Hobson, K. A. 2003. Making migratory connections with stable isotopes. Pages 379–391. *in* P. Berthold, E. Gwinner, and E. Sonnenschein (Eds.) *Avian Migration.* Springer-Verlag, Berlin Heidelberg, New York.

Hobson, K. A. 2005. Stable isotopes and the determination of avian migratory connectivity and seasonal interactions. *Auk* **122**:1037–1048.

Hobson, K. A., and L. I. Wassenaar. 1997. Linking breeding and wintering grounds of neotropical migrant songbirds using stable hydrogen isotopic analysis of feathers. *Oecologia* **109**:142–148.

Holland, R. A., K. Thorup, M. J. Vonhof, W. W. Cochran, and M. Wikelski. 2006. Bat orientation using Earth's magnetic field. *Nature* **445**:702.

Janssens, E., T. Dauwe, L. Bervoets, and M. Eens. 2002. Inter- and intraclutch variability in heavy metals in feathers of Great Tit nestlings (*Parus major*) along a pollution gradient. *Archives of Environmental Contaminants and Toxicology* **43**:323–329.

Jouventin, P., and H. Weimerskirch. 1990. Satellite tracking of Wandering Albatrosses. *Nature* **343**:746–748.

Kaminski, R. M., and E. A. Gleusing. 1987. Density and habitat related recruitment in mallards. *Journal of Wildlife Management* **51**:141–148.

Karasov, W. H., and C. Martinez del Rio. 2007. Isotope ecology. Pages 433–478 *in Physiological Ecology*. Princeton University Press, Princeton, New Jersey.

Kelly, J. F., V. Atudorei, Z. D. Sharp, and D. M. Finch. 2002. Insights into Wilson's Warbler migration from analyses of hydrogen stable-isotope ratios. *Oecologia* **130**:216–221.

Kelly, J. F., K. C. Ruegg, and T. B. Smith. 2005. Combining isotopic and genetic markers to identify breeding origins of migrant birds. *Ecological Applications* **15**:1487–1494.

Kelsall, J. P. 1984. The use of chemical profiles from feathers to determine the origins of birds. Pages 501–515 *in* J. Ledger (Ed.) *Proceedings of the fifth Pan-African Ornithological Congress, Lilongwe, Malai, 1980*. South African Ornithological Society, Johannesburg.

Lajtha, K., and R. H. Michener (Eds.). 1994. *Stable Isotopes in Ecology and Environmental Science*. Blackwell Scientific Publications, Oxford.

Levey, D., and F. G. Stiles. 1992. Evolutionary precursors of long-distance migration: Resource availability and movement patterns in neotropical landbirds. *American Naturalist* **140**:447–476.

Liu, J., H. Xiao, F. Lei, Q. Zhu, K. Qin, X.-W. Zhang, X.-L. Zhang, D. Zhao, G. Wang, Y. Feng, J. Ma, W. Liu, *et al.* 2005. Highly pathogenic H5N1 influenza virus infection in migratory birds. *Science* **309**:1206–000.

Marra, P. P., K. A. Hobson, and R. T. Holmes. 1998. Linking winter and summer events in a migratory bird by using stable-carbon isotopes. *Science* **282**:1884–1886.

Marra, P. P., D. R. Norris, S. M. Haig, M. S. Webster, and J. A. Royle. 2006. Migratory connectivity. Pages 157–183. *in* K. R. Crooks and M. A. Sanjayan (Eds.) *Connectivity Conservation*. Cambridge University Press, New York.

Martin, T. M., I. Chades, P. Arcese, P. P. Marra, H. P. Possingham, and D. R. Norris. 2007. Optimal conservation of migratory birds. *Public Library of Science, ONE* **2**(8):e571.

Michener, R. M., and K. Lajtha (eds.). 2007. Stable Isotopes in Ecology and Environmental Science, Second Edition. Blackwell Publishing, Oxford.

Møller, A. P., and W. Fiedler. 2004. Birds and climate change., P. Berthold (Ed.) *In Advances in Ecological Research*, Vol. 35. Elsevier, Oxford, UK.

Moore, F. R., R. J. Smith, and R. Sandberg. 2005. Stopover ecology and intercontinental migrants: En route problems and consequences for reproductive performance. Pages 251–261. *in* R. Greenberg and P. Marra (Eds.) *Birds of Two Worlds—the Ecology and Evolution of Migration*. Johns Hopkins University Press, Baltimore, Maryland.

Murray, M. R., and D. L. Fuller. 2000. A critical review of the effects of marking on the biology of vertebrates. Pages 15–64 in Research Techniques in Animal Ecology: Controversies and Consequences. Methods and Cases in Conservation Science (L. Boitani and T. K. Fuller, eds.). Columbia University Press, New York.

Newton, I. 2006. Advances in the study of irruptive migration. *Ardea* **94**:433–460.

Nolan, E. D., and V. Ketterson. 1976. Geographic variation and its climatic correlates in the sex ratio of eastern-wintering dark-eyed juncos (*Junco hyemalis hyemalis*). *Ecology* **57**:679–693.

Norris, D. R., and P. P. Marra. 2007. Seasonal interactions, habitat quality and population dynamics in migratory birds. *Condor* **109**:535–547.

Norris, D. R., P. P. Marra, T. K. Kyser, T. W. Sherry, and L. M. Ratcliffe. 2004. Tropical winter habitat limits reproductive success on the temperate breeding grounds in a migratory bird. *Proceedings of the Royal Society of London Series B* **271**:59–64.

Norris, D. R., P. P. Marra, T. K. Kyser, J. A. Royle, G. J. Bowen, and L. M. Ratcliffe. 2006. Migratory connectivity of a widely distributed Neotropical-Nearctic migratory songbird. *Ornithological Monographs* **61**:14–28.

Norris, D. R., D. B. Lank, J. Pither, D. Chipley, R. C. Ydenberg, and T. K. Kyser. 2007. Trace elements identify wintering populations of a migratory shorebird. *Canadian Journal of Zoology* **85**:579–583.

Ochoas-Acuna, H., M. S. Sepulveda, and T. S. Gross. 2002. Mercury in feathers from Chilean birds: Influence of location, feeding strategy, and taxonomic affiliation. *Marine Pollution Bulletin* **44**:340–349.

Owen, M., and J. M. Black. 1989. Factors affecting the survival of Barnacle Geese on migration from the breeding grounds. *Journal of Animal Ecology* **58**:603–617.

Parrish, J. R., D. T. Rogers Jr., and F. P. Ward. 1983. Identification of natal locales of Peregrine Falcons (*Falco peregrinus*) by trace element analysis of feathers. *Auk* **100**:560–567.

Perrins, C. M. 1970. Timing of birds breeding seasons. *Ibis* **112**:242–255.

Pliny, the Elder. 1855. *The Natural History.* John Bostock, H.T. Riley (Trans.) Taylor and Francis, Red Lion Court, Fleet Street.

Pulido, F. 2007. The genetics and evolution of avian migration. *BioScience* **57**:165–174.

Quinn, T. P. 2005. *The Behavior and Ecology and Pacific Salmon and Trout.* American Fisheries Society, Bethesda Maryland and the University of Washington Press, Seattle.

Ramenofsky, M., and J. C. Wingfield. 2007. Regulation of migration. *BioScience* **57**:135–143.

Ramos, M. A. 1983. *Seasonal Movements of Bird Populations at a Neotropical Study Site in Southern Veracruz, Mexico.* Ph.D. dissertation. University of Minnesota, Minneapolis.

Ramos, M. A., and D. W. Warner. 1980. Analysis of North American subspecies of migrant birds wintering in Los Tuxtlas, southern Veracruz, Mexico. Pages 172–180 *in* A. Keast and E. S. Morton (Eds.) *Migrant Birds in the Neotropics: Ecology, Behavior, Distribution, and Conservation.* Smithsonian Institution Press, Washington, DC.

Rappole, J. H., S. R. Derrickson, and Z. Hubalek. 2000. Migratory birds and the spread of West Nile virus in the western hemisphere. *Emerging Infectious Diseases* **6**:319–328.

Reed, K. D., J. K. Meece, J. S. Henkel, and S. K. Shukla. 2003. Birds, migration and emerging zoonoses: West Nile Virus, Lyme Disease, Influenza A and enteropathogens. *Clinical and Medical Research* **1**:5–12.

Ricklefs, R. E., S. M. Fallon, S. C. Latta, B. L. Swanson, and E. Bermingham. 2005. Migrants and their parasites: A bridge between two worlds. Pages 210–221 *in* R. Grrenberg and P. MarraM (Eds.) *Birds of Two Worlds.* Johns Hopkins University Press, Baltimore.

Riley, J. R., A. D. Smith, D. R. Reynolds, A. S. Edwards, J. L. Osborne, I. H. Williams, N. L. Carreck, and G. M. Poppy. 1996. Tracking bees with harmonic radar. *Nature* **379**:27–30.

Robinson, S. K., F. R. Thompson, T. M. Donovan, D. R. Whitehead, and J. Faaborg. 1995. Regional forest fragmentation and nesting success of migratory birds. *Science* **267**:1987–1990.

Roff, D. A., and D. J. Fairbairn. 2007. The evolution and genetics of migration in insects. *BioScience* **57**:155–164.

Rubenstein, D. R., and K. A. Hobson. 2004. From birds to butterflies: Animal movement and stable isotopes. *Trends in Ecology and Evolution* **19**:256–263.

Rundel, P. W., J. R. Ehleringer, and K. A. Nagy. 1988. *Stable Isotopes in Ecological Research.* Springer-Verlag, Berlin.

Salomonsen, F. 1955. The evolutionary significance of bird migration. *Biologiske Meddelesler* **22**:1–62.

Sanz, J. J., J. Potti, J. Moreno, S. Merino, and O. Frias. 2003. Climate change and fitness components of a migratory bird breeding in the Mediterranean region. *Global Change Biology* **9**:461–472.

Shaffer, S. A., Y. Tremblay, J. A. Awkerman, R. W. Henry, S. L. H. Teo, D. J. Anderson, D. A. Croll, B. A. Block, and D. P. Costa. 2005. Comparison of light- and SST-based geolocation with satellite telemetry in free-ranging albatrosses. *Marine Biology* **147**:833–843.

Shaffer, S. A., Y. Tremblay, H. Weimerskirch, D. Scott, D. R. Thompson, P. M. Sager, H. Moller, G. A. Taylor, D. G. Foley, B. A. Block, and D. P. Costa. 2006. Migratory shearwaters integrate oceanic resources across the Pacific Ocean in an endless summer. *Proceedings of the National Academy of Sciences of the United States of America* **103:**12799–12802.

Sibley, D. A. 2001. *National Audubon Society Sibley Guide to Birds*. Alfred A. Knopf, New York.

Smith, T. B., S. M. Clegg, M. Kimura, K. C. Ruegg, B. Mila, and I. Lovette. 2005. Molecular and genetic approaches to linking breeding and wintering areas in five Neotropical migrant passerines. Pages 222–234 *in* R. Grrenberg and P. Marra (Eds.) *Birds of Two Worlds*. Johns Hopkins University Press, Baltimore.

Stokely, J. M. 2005. *The Feasibility of Utilizing the Cellular Infrastructure for Urban Wildlife Telemetry.* Ph.D. dissertation. Virginia Polytechnical Institute and Virginia State University.

Szep, T., A. Moller, J. Vallner, B. Kovacs, and D. Norman. 2003. Use of trace elements in feathers of sand martin *Riparia riparia* for identifying moulting areas. *Journal of Avian Biology* **34:**307–320.

Vega-Rivera, J. H., J. H. Rappole, W. J. McShea, and C. A. Haas. 1998. Wood Thrush postfledging movements and habitat use in northern Virginia. *Condor* **100:**69–78.

Visser, M. E., A. J. van Noordwijk, J. M. Tinbergen, and C. M. Lessells. 1998. Warmer springs lead to mistimed reproduction in great tits (*Parus Major*). *Proceedings of the Royal Society of London Series B—Biological Sciences* **265:**1867–1870.

Visser, M. E., C. Both, and M. M. Lambrechts. 2004. Global climate change leads to mistimed avian reproduction. *Advances in Ecological Research* **35:**89–110.

Webster, M. S., and P. P. Marra. 2005. The importance of understanding migratory connectivity. Pages 199–209 *in* R. Greenberg and P. P. Marra (Eds.) *Birds of Two Worlds: The Ecology and Evolution of Temperate-Tropical Migration Systems*. Johns Hopkins University Press, Baltimore, Maryland.

Webster, M. S., P. P. Marra, S. M. Haig, S. Bensch, and R. T. Holmes. 2002. Links between worlds: Unraveling migratory connectivity. *Trends in Ecology and Evolution* **17:**76–83.

Winkler, D. W., P. O. Dunn, and C. E. McCulloch. 2002. Predicting the effects of climate change on avian life-history traits. *Proceedings of the National Academy of Sciences* **99:**13595–13599.

Wikelski, M., E. M. Tarlow, A. Raim, R. H. Diehl, R. P. Larkin, and G. H. Visser. 2003. Costs of migration in free-flying songbirds. *Nature* **423:**704.

Wikelski, M., D. Moskowitz, J. S. Adelman, J. Cochran, D. S. Wilcove, and M. L. May. 2006. Simple rules guide dragonfly migration. *Biology Letters* **2:**325–329.

Wikelski, M., R. W. Kays, N. Jeremy Kasdin, K. Thorup, J. A. Swenson, and G. W. Smith Jr. 2007. Going wild: What a global small-animal tracking system could do for experimental biologists. *Journal of Experimental Biology* **210:**181–186.

CHAPTER 2

An Introduction to Light Stable Isotopes for Use in Terrestrial Animal Migration Studies

Leonard I. Wassenaar

Environment Canada

Contents

I. INTRODUCTION

Stable isotopes of the light elements (C, H, N, O, and S) are increasingly employed in terrestrial animal migration studies. The primary reason is that isotopes are powerful forensic recorders of dietary sources that can be spatially interpolated or explicitly linked to on-the-ground and large-scale patterns in the landscape and hydrosphere at a variety of scales (Hobson and Wassenaar 1997,

Tracking Animal Migration with Stable Isotopes
K. A. Hobson and L. I. Wassenaar (Editors)
ISSN 1936-7961, DOI: 10.1016/S1936-7961(07)00002-4

Bowen *et al.* 2005b). Depending on whether the tissue of the migrating animal is biochemically fixed (feather, hair) or dynamic (blood, muscle), these isotopic dietary tracers record fundamental information about where an organism is and what it has been eating (Chapter 3). This salient feature opens a forensic door that allows us to gather new insights into one of the most confounding aspects of animal migration—their mobility.

There are stunningly few scientific tools available at our disposal to facilitate and quantify the tracking and movement of small animals over variable or large uninhabited spatial distances. In addition, the spatial scale of migration is often continental to global in scope. The focus of this volume is on one of the more newly available tools in the researchers toolbox—light stable isotope methods. Stable isotopes are not to be misconstrued as a "silver bullet" to be indiscriminately applied to answer all animal migration questions. Scientists, in practice, seek to carefully employ as many tools as possible to answer a specific research question, and one of these tools may include stable isotope methods. Further, it should be emphasized that stable isotopic investigations into the various kinds of terrestrial (and aquatic) faunal migration phenomena are relatively recent (Hobson 1999), and the accumulating literature is clear that isotopic approaches are evolving, methodologies are improving, and interpretational challenges are being actively identified and addressed. While stable isotope approaches have already illustrated remarkable successes in generating new knowledge and insight into animal migration, the limits of stable isotope tracing of migrants (from basic measurement to spatial interpretations) remain to be fully probed and tested.

Briefly, the scientific tools currently available for tracing the migratory movement of terrestrial biota fall into two general categories—*extrinsic* and *intrinsic markers* (see Chapter 1 for full listing). Stable isotopes fall into the category of intrinsic markers of dietary and spatial origin. Parlayed into everyday terms, stable isotopes enable us to determine where our migrant had its lunch by our direct or inferred knowledge of geospatial differences in the lunch menu.

Stable isotopes as intrinsic markers provide several key advantages over existing extrinsic markers. Principally, intrinsic markers do not require a subsequent recapture of the same organism to obtain a successful result because the required spatial data is recorded within each organism's tissue. As noted in Chapter 1, the success of physical tags is generally poor, and extrinsic markers are usually not feasible for small bodied migrants such as insects or small birds. Similarly, radio or satellite telemetry is technologically currently restricted to larger migrant organisms and requires subsequent electronic contact—a form of recapture. External tags and markers may also unwittingly alter or hinder normal movement behavior until miniaturization is improved.

Another comparative advantage of intrinsic markers over extrinsic markers is that they are not inherently spatially biased. A drawback to the more popular extrinsic markers (*e.g.*, tags, rings) is that the results are proportionally biased to effort at the location of marking, be it at monitoring sites, staging areas, areas within existing observational networks, or field accessibility. While this bias may not be problematic for local population interaction studies, it remains a major hurdle to our understanding of large distance animal migration. For some species of interest, extrinsic markers exclude important data from large unmonitored portions of the migratory catchment areas used by migrants (*e.g.*, boreal, far north). Intrinsic markers, therefore, hold both the potential to extend our knowledge into spatial areas where extrinsic markers are virtually impossible to deploy.

In order for light stable isotopes to function as intrinsic markers of migrant mobility, some key prerequisites must be met. The foremost, and rather obvious, requirement is that the migrant organism contains one or more of the light isotopes of interest in specific tissues. Fortunately, the light isotopes form the key atomic building blocks of the biosphere and most animal tissues; therefore, this requirement is handily met (Table 2.1). The second prerequisite is that the organism migrates between isotopically distinct landscapes or biomes (*e.g.*, Chapter 4) and retains in one or more of its tissues, permanently (*e.g.*, feather) or integrated over some period of time (*e.g.*, muscle), measurable isotopic differences that can be linked to diet at previous or current locations. This requirement is most easily met for species with discrete spatial distribution ranges that migrate seasonally across distinct isotopic landscapes.

TABLE 2.1	The approximate elemental abundances as dry weight %, stable isotope ratios of interest, and generalized isotopic ranges for bulk tissues (*e.g.*, α- or β-keratins like hair or feathers) commonly used in migratory research

Element	Weight	Isotope ratios	δ Range (‰)	Mass required (mg)
Light Isotopes				
Carbon[a]	30–40 wt.%	$^{13}C/^{12}C$	−5 to −65 (PDB)	0.2–1.5
Oxygen[b]	27–40 wt.%	$^{18}O/^{16}O$	+10 to +30 (VSMOW)	0.2–0.5
Nitrogen[a]	12–19 wt.%	$^{15}N/^{14}N$	−2 to +25 (Air)	0.5–1.5
Hydrogen[b]	6–8 wt.%	$^{2}H/^{1}H$	−250 to +90 (VSMOW)	0.1–0.4
Sulfur	5–20 wt.%	$^{34}S/^{32}S$	−20 to +30 (CDT)	1–2
Heavy Isotopes				
Strontium	<100-? ppb	$^{87}Sr/^{86}Sr$	Absolute ratios	2–30[c]

[a] Carbon and Nitrogen isotopes are generally obtained simultaneously by CF-IRMS, but need sufficient mass to obtain enough N.
[b] Hydrogen and Oxygen may be obtained simultaneously by pyrolytic methods.
[c] Dependent on the Strontium concentrations previously determined.
Keratins commonly contain the smaller amino acids such as glycine, alanine, and cysteine. Mass of sample typically required for each isotopic assay also applies to most other biological tissues like muscle, claw, and blood. The primary isotopic standards are PeeDee Belemnite (PDB), Vienna Standard Mean Ocean Water (VSMOW), atmospheric N_2 (AIR), and Canyon Diablo Triolite (CDT).

One example would be redhead ducks that migrate between higher latitude northern boreal forests and lower latitude estuarine areas of the coastal Caribbean (Hobson *et al.* 2004a). A spatial isotopic linkage also requires that the isotopic offsets (net isotope discrimination) between these seasonally used dietary landscapes and the selected tissue being measured is consistent or well known, or has been empirically measured. Of course, such ideal conditions are often not realized and may be complicated by a host of issues that will provide research opportunities and challenges for many years to come.

Most biologists and ecologists that conduct avian or terrestrial insect migration research projects do not operate a costly stable isotope laboratory facility, and so are reliant on specialized commercial, government, or university stable isotope laboratories that can analyze their samples for them on a fee-for-service basis, or by a collaborative project. This interaction will lead to queries and discussions between an isotopic analyst who may have little or no knowledge of the migration project at hand, and a biologist or ecologist who may have little or no knowledge about the intricacies of stable isotope jargon and the required analytical protocols and procedures for a successful analysis. Because of the potential for confusion or unrealistic expectations, and because stable isotope analyses may be a costly budget item, the intent of this chapter is to help bridge a discipline gap by arming the ecologist with the essential information required to ensure a good understanding of the terminology and to gain confidence that the kinds of isotopic assays sought are appropriate and useful. Practically, the types and amount of sample matrix or tissues that can be analyzed for the key light isotopes will be covered, because much of the analytical budget may be overtaken by preparative procedures that are easily done in the ecologists' laboratory by students and support staff.

From experience, a common scenario starts with a query from a researcher or graduate student who is interested in the migration ecology of a particular species. They have read published papers or viewed a presentation at a scientific meeting suggesting that stable isotopes can be used track animal migration. What follows is a dialogue among colleagues that form the outline of this chapter. What are stable isotopes? Which stable isotopes will be of use to the migration research questions for my species of interest? What sample or tissue is needed for the analysis? How do I prepare the sample? How much sample is required? Are there caveats? What kind of analytical error can I expect? How much will it cost? Why do my duplicates give different results? These are but a few of the fundamental questions and concerns that, if addressed as fully as possible from the start, will form the basis of a well-designed successful migration research project using stable isotopes.

The goal of this chapter, therefore, is to provide a practical guide for ecologists and biologists wishing to employ light stable isotopes as intrinsic markers in their terrestrial animal migration research projects. While the overall thrust of this volume is primarily on larger- and continental-scale movement of animals because of the longstanding and intractable problems associated with documenting this massive scale of movement, we must acknowledge that the scale of animal migration can be tiny indeed. One example may be the seasonal vertical migration of earthworms in soils of less than a meter. The focus here is to a give concise and practical guidance to the nonisotope specialist on obtaining the best possible stable isotope data in order to ensure that the assays are appropriately done, and as a result meaningful spatial interpretations regarding migratory movement can be made.

A. What Are the Light Stable Isotopes?

Most of the elements of the periodic table have a number of stable isotopes, that is, elements having the same number of protons but differing only in their numbers of neutrons (Criss 1999, Hoefs 2004). However, of the many elements that have stable isotopes, there are just a handful that are of immediate practical interest to studies of animal migration and ecology. These are the so-called "light isotopes" of the elements of C, N, H, O, and S. These five atomic elements comprise the primary building blocks of all components of the biosphere (plants, animals), the hydrosphere (water), and the atmosphere (gaseous N_2, O_2, H_2O). For example, we can see that these five elements, to varying proportions, overwhelmingly comprise the bulk of all animal tissues, from $\sim$50% dry weight for C to lower levels of 5–6% dry weight for H (Table 2.1). These five elements each have two or more stable isotopes that vary widely in nature from the micro- to macroenvironmental scale. Other "heavier" elements and their isotopes may also be useful but tend to occur at trace concentrations and do not form the majority of organic tissue structure. One example of a "heavy isotope" trace element that can be extracted from tissue and that has spatial distribution patterns is the element strontium (Sr, Table 2.1).

All of the light isotopes have a common or abundant "light" isotope (*e.g.*, ^{12}C; 98.894%) and one or more "heavier" rare isotope of interest (*e.g.*, ^{13}C; 1.1056%; Hoefs 2004). It is the abundance ratios of these light and heavy isotopes that vary minutely in nature because of physical and chemical processes, and these variations are ultimately of interest to us. However, it is exceedingly difficult to measure absolute ratios with any great accuracy, or to determine absolute concentrations of each isotope in a sample. It was recognized nearly 60 years ago, however, that it is easier to precisely measure *relative* differences in the isotopic ratios between a pure gaseous sample (*e.g.*, CO_2, H_2, SO_2, N_2) and gaseous reference using a gas source mass spectrometer (McKinney *et al.* 1950).

The fact that *gas source* isotope ratio mass spectrometers (IRMSs) are primarily used to measure the light isotopes has two important implications. First, it means that all samples (feather, muscle, blood, claw, etc) cannot have their isotope ratios measured directly on the raw organic material or sample provided by the researcher, but the sample must first be combusted or by some means quantitatively converted to an ultrapure and appropriate gaseous analyte in order to measure its isotopic ratios relative to a calibrated reference gas of the same type. Numerous papers and books have been devoted to preparative conversions for a host of organic substrates over five decades—the most recent summaries and historical perspectives on this topic are in Volumes 1 and 2 of the *Handbook of Stable Isotope Techniques* (de Groot 2004).

Most typically for organic solid substrates, samples are quantitatively combusted using elemental analyzers (EAs; Figure 2.1) to convert samples to pure CO_2, H_2O, SO_2, or N_2 gases (Fry *et al.* 1992). While the fundamental principles of the IRMS remain virtually unchanged since the 1950s, much of the developmental in the last decade have focused on automating sample preparation in order to obtain multiple isotope assays from a single sample or new substrate, on decreasing sample size requirements, or on improving sample throughput rates while maintaining highest measurement accuracy.

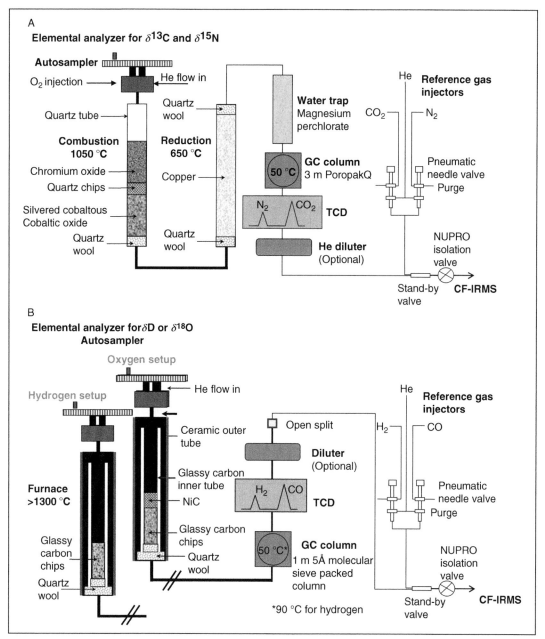

FIGURE 2.1 Typical elemental analyzer (EA) preparative interface configurations to a CF-IRMS system for (A) $\delta^{13}C$ and $\delta^{15}N$ and (B) δD or $\delta^{18}O$ in organic tissue samples (courtesy of GV Instruments).

The second implication is that stable isotopic ratio results reported back to the researcher are not provided in a readily recognized SI concentration format such as mg/Liter or μmol/g. Instead the researcher will receive a series of "δ" numbers having positive or negative values expressed as a parts per thousand (‰) relative difference to an international standard (Hoefs 2004, Fry 2006). For the uninitiated this can be confusing. But recall, because we can very precisely measure minute relative

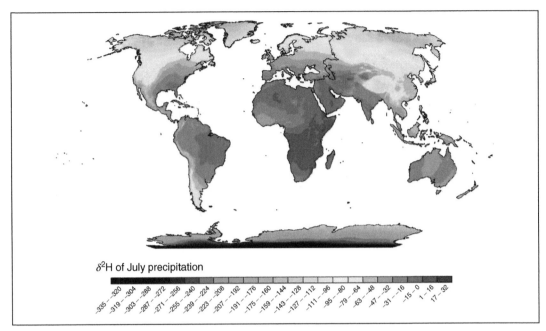

δ^2H of July precipitation

FIGURE 2.2 The average hydrogen isotopic (δD) composition of rainfall across the globe in the summertime. These large-scale and distinct isotope patterns are largely mirrored in plants and into upper-level trophic organisms and so form the basis for tracking animal movement. Vectors of seasonal migratory movement are often North-South and spanning large H isotopic gradients. The terrestrial global range of δD is over 300‰, with a measurement precision of better than ±2‰. The oxygen isotopes form a similar pattern. The oceans are isotopically homogenous at ~0‰ for both δD and δ^{18}O. Map courtesy of G. Bowen and may be found online at www.wateristotopes.org. (**See Color Plate.**)

isotopic differences between two gases, an examination of the standard "δ" equation reveals what the δ data mean:

$$\delta X(‰) = \left(\frac{\text{Isotopic ratio}_{\text{sample}}}{\text{Isotopic ratio}_{\text{standard}-1}} \right) \times 1000 \qquad (2.1)$$

where X on the left side is the isotopic element of interest (δ^{13}C, δ^{18}O, δD, etc). The right side of the equation in parentheses is the isotopic ratios of the sample and international standard comparatively measured by the mass spectrometer in the laboratory. The isotopic ratio is the measured ratio of the light to heavy isotopes (^{13}C/^{12}C, ^{18}O/^{16}O). The measured ratio results are taken and multiplied by 1000 simply for convenience in order to transform the values primarily into whole numbers (*e.g.*, easier to work with than 5–6 decimal place numbers). Because the light isotope primary references were established arbitrarily many decades ago (these are our "zero" points), the results can be negative or positive values in parts per thousand (‰) deviations because they are always *relative to the accepted international standard* (Groning 2004). Thus, we need never be concerned about the positive or negative sign. The primary stable isotope reference materials currently used are listed in Table 2.1. Because primary references are preciously limited in quantity, most laboratories routinely employ local laboratory standards that have been carefully calibrated against primary reference materials.

For example, a sample reported having a δD value of +150‰ means the sample has 150 parts per thousand (or 15%) more deuterium in it than the standard, in this case the Vienna Standard Mean Ocean Water (VSMOW) standard. If the value were negative, it simply means 150 parts per thousand less deuterium in it than VSMOW. While these isotopic concentration differences are amazingly small, it is worth noting that modern IRMSs can reliably measure isotopic ratio *differences* to the sixth decimal place, or expressed in ‰, to ±0.01‰, or better. The limiting factor, and indeed most of the analytical error, generally occurs in the preparative conversions of raw sample to the appropriate analyte gas of interest or in the sample gas handling and transfer systems.

A final term that must be briefly explained is mass-dependent *isotope fractionation*. Isotope fractionation occurs when a chemical reaction or a process results in a changing of the stable isotope ratios of the source or reactant because of the slight chemical differences arising from the subtle differences in mass. Unidirectional or irreversible isotope fractionations are known as *kinetic isotope fractionations*, whereas *equilibrium fractionations* (*e.g.*, $^{18O:16O}O_{2\ gas} \leftrightarrow {}^{18O:16O}O_{2\ dissolved}$) are chemical reactions that are reversible (Hoefs 2004). Sometimes isotope fractionation is less explicitly portrayed as "δ differences" between two or more partially related substrates. For example, in this volume, the emphasis on hydrogen isotopes as a tracer of migration is key, whereby patterns in the hydrosphere (water) are linked to those in plants and animals. While there may be consistent δ "offsets" among these three hydrogen-bearing components at each geospatial site, the water-diet-tissue "fractionations" is more accurately the "net isotopic discriminations" because every single hydrogen pool and form in those three substrates are not measured or fully known, nor do we follow all individual hydrogen atoms, but instead rely on selected analyses of bulk samples. The reader is referred to the textbooks below for exhaustive discussion of the various types of isotope fractionation occurring in nature. In short, were there no isotope fractionation in nature, all components of the geosphere, hydrosphere, atmosphere, and biosphere would have the exact same isotopic ratios and stable isotopic assays would be pointless. Fortunately, isotopic fractionations of the light isotopes in nature are widespread, diverse, and often characteristic in their magnitude and direction. This feature allows us to exploit isotopic measurements for the purposes of studying animal migration (Chapters 3–5).

The type of modern gas IRMS overwhelmingly used for biological and migration studies these days is called *continuous-flow isotope ratio mass spectrometry* (CF-IRMS), which gained widespread acceptance in the 1990s, and has been the subject of intense development since the late 1980s (Matthews and Hayes 1978). Compared to the highest precision *dual-inlet isotope ratio mass spectrometry* (DI-IRMS) developed in the 1950s, there are pros and cons to CF-IRMS assays. The primary and most attractive aspects of CF-IRMS includes much lower per-sample analytical cost, a high degree of automation by linking preparative modules such as EAs, and high sample throughput rates. The main disadvantage remains lower analytical precision, although in recent years, many CF-IRMS assays for CNHOS approach or exceed dual inlet assays simply due to automation and improvements in sample preparative handling. ^{34}S is still best done by dual inlet for the highest precision, although CF assays are constantly improving. Radiogenic Sr isotopes are conducted by solid source IRMS. The cost in US dollars of stable isotope analyses in 2007 range from as low as $10–15 for $^{13}C + {}^{15}N$ to $15–50 for $^{18}O/^{2}H$, $50–100 for ^{34}S to $100 or more for $\delta^{87}Sr$. Schematic examples of some typical CF-IRMS preparative EA systems for C, N, H, and O isotope analyses are shown in Figure 2.1.

It is beyond the scope of this chapter to exhaustively cover all aspects of stable isotope mass spectrometry and biogeochemistry. The reader is referred to some key textbooks published over the last decade that cover in detail the scope of stable isotope theory and environmental applications (Kendall and McDonnell 1998, Criss 1999, Hoefs 2004). For students and newcomers to the field of stable isotopes, a good place to start for fundamental introduction and discussion of stable isotopes in the biosphere and hydrosphere are *Stable Isotope Ecology* (Fry 2006), *Stable Isotopes in Ecology and Environmental Science* (Lajtha and Michener 2007), *Stable Isotope Geochemistry* (Sharp 2007), and *Environmental Isotopes in Hydrogeology* (Clark and Fritz 1997).

II. MATERIALS AND METHODS

In the following two sections, we will cover field sample collection and tissue subsampling and discuss each of the light isotopes, grouped in order of the type of information that can be gained that may be translated into spatial analyses useful for migratory research.

For the purposes of terrestrial animal migration research, stable H and O isotopes in the terrestrial environment may be considered "*global-spatial*" assays because the patterns of H and O isotopes on terrestrial systems are systematically controlled by global-scale hydrologic and meteorological processes that are seasonally and spatially predictable over multiyear time frames and continuously over massive geospatial scales (regional, continental, global). This salient feature provides the highest level of confidence in making spatial interpolations into areas where no long-term data or stations exist (Figure 2.2, Bowen *et al.* 2005b). Note that the oceans are global O and H reservoirs and are largely isotopically homogenous, which further allows clear marine versus terrestrial distinctions to be made.

C, N, S, and Sr isotopes may be grouped into "*local-spatial*" assays, mainly because there are no strong *a priori* reasons to predict that these isotopes vary systematically and continuously on the landscape or over very large geospatial scales (although they *could* pattern over smaller scales) to the same extent as H and O. While there are well-known spatial isotopic differences in key isotopes (*e.g.*, ^{13}C differences between C_3 and C_4 plant-dominated habitats), local variables may be strongly and unpredictably influential for C, N, and S, and for Sr include soil type, altitude, agricultural land use, industrial influences, and local geology (Hebert and Wassenaar 2001). These local-scale variations may preclude or hinder predictive continuous spatial interpolations. Ongoing and new research suggests that some larger-scale patterns in ^{13}C and ^{15}N may occur (*e.g.*, terrestrial vs marine), but this area requires further development (Chapters 4). The primary controls on Sr isotope ratios are local geology (spatially variable or discontinuous) and the ocean (isotopically homogeneous) that allows for clear marine versus terrestrial distinctions, but also terrestrial fingerprinting.

The first approach to large-scale animal migration investigations using intrinsic tracers like stable isotopes is likely to be the most fruitful by using the global-spatial assays (H and O isotopes). However, spatial resolving power and details of habitat use may be greatly improved by careful consideration and selection of a subset of local-spatial isotopes (C, N, S, and Sr) or by including other assays and data (*e.g.*, extrinsic data) and a Bayesian analysis approach. Discussions on this and multi-isotope spatial assignment approaches (and difficulties) are outlined in detail in Chapters 4 and 5 and elsewhere (Hebert and Wassenaar 2005).

A. Sample Collection and Preparation

One of the very first questions encountered by researchers using stable isotope methods for migration research involves the type of samples and subsamples that need to be collected from the organism of interest. This could be hair, claw, muscle, blood, wings, thorax, etc. There are two main categories of samples that may be defined: (1) *fixed tissues* and (2) *dynamic tissues*.

Fixed tissues are here defined as those kinds of discrete tissues or body parts that once formed are metabolically (and isotopically) inert, recording the isotopic composition of local diet at or close to the time of formation. These include the so-called "dead" tissues such as keratinous claw, nails, hair, and feathers. One of the primary and most often used fixed tissues are bird feathers—once fully formed, the vane material and rachis (stem) does not change chemically or isotopically as the organism moves away from the location of formation and along its migratory route. Key advantage of feathers is that for many species they are often grown over a very short period of time, or at one specific location. For many birds and insects, feathers or wings are often formed at the breeding site, wintering ground, or site of emergence, and so will record the isotopes of diet at that specific location. Fixed tissues are most often used for estimating net spatial vectors from the point of origin of formation.

However, there are some caveats to the use of fixed tissues that require the researcher to have a good knowledge of the ecology and physiology of the organism under study. Potential complications include slow or variably growing fixed tissues that can lead to the issue of intra- or intersample isotopic variance arising from transient dietary conditions during movement. As an illustrative example, consider human hair. Human hair forms at an average rate of ~10–15 mm per month depending on the location on the scalp, but the hair is isotopically "fixed" once formed and therefore records information about the location of diet at the time of cellular growth. There may be some lag time before local diet is isotopically recorded in the hair, but a temporal record of diet along the length of the hair sample can be investigated and spatially linked (Hobson and Schell 1998, O'Brien and Wooller 2007).

Another cautionary example involves slowly growing feathers, claws, or hairs that form as an animal is enroute and ingesting diet at different geospatial locations and stopovers. This is aptly illustrated in Figure 2.3 showing hydrogen isotope data obtained along the length of a single flight feather from a bald eagle (see also intrasample issues below). The data clearly revealed distinct hydrogen isotopic patterns along the entire length of a single flight feather that can be related to southern migratory movement in North America *during* feather growth. This bald eagle had been tagged and began its flight feather growth and migratory journey in Saskatchewan, Canada (*e.g.*, the more negative δD values in older feather material near the tip) and then migrated southward into the United States during the fall (positive δD values toward the base), and was recaptured the following year in Canada long after the feather was fully grown. It is clear that as the feather was temporally extruded from the base, it gained more positive δD values as the eagle moved to more southerly locations, and essentially was recording a "net migratory track" within a single feather. However, had one not known the feather was grown enroute and had taken random subsamples and assumed they were solely indicative of natal

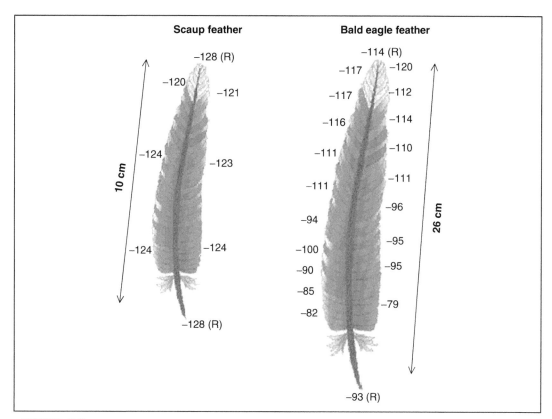

FIGURE 2.3 Illustration of measured intrasample hydrogen (δD) isotopic heterogeneity in migrating birds. Illustrated is a comparison of a flight feather from lesser scaup (*Aythya affinis*), left, that grew the entire feather at one location and a bald eagle (*Haliaetus leucocephalus*), right, that grew the feather along its southward migration. Subsamples for hydrogen isotope analyses were taken from vane material (350 μg) up and down the left and right side of a single feather, and from the top and bottom portions of the rachis (R). Note that illustrated differences in feather length are not to scale and are in centimeters from the tip to base. δD results are in percentages (VSMOW).

origin, one could have obtained highly confounding results from this individual and among eagles. Further, movement during tissue growth is another primary reason why duplicate samples taken from different parts of the same feather or tissue (often assumed by the researcher to be true repeats) can give startlingly different results. In other words, what may have been presumed to be a single fixed tissue sample (a feather) may in fact be a potentially interesting spatial "recorder."

By contrast, the intrafeather hydrogen isotope patterns from a migrant lesser scaup (*Athya affinis*) are shown in Figure 2.3. For this individual, it is known the entire flight feather was formed at the natal location in Canada. Here no significant intrasample tissue isotopic differences were observed that would indicate that this individual had moved during feather formation, despite this migrant being captured far from its natal site. In other words, this feather is an excellent candidate to assess postbreeding spatial vectors, regardless of the position on the feather where the subsample for isotope assays was taken.

Other confounding aspects of sampling may include partial molts or feather loss, differential molt times (head feathers vs breast vs flight feathers at different times and locations), or dietary switches. For fixed tissues on newborn or maternally fed animals, there is also the possibility that some of the isotopic information is obtained from the mother but was derived at another location (feeding far from the natal site or from preexisting body reserves obtained elsewhere) (Duxbury *et al.* 2003).

Another category of fixed or dynamic tissues are those preserved in museum or archaeological collections. Such collections may be attractive in order to gain historical trends information regarding migrating or even from extinct species. However, these fixed tissues may have be been preserved (*e.g.*, formalin storage), chemically treated, or dried. The researcher must therefore be vigilant to the potential isotope fractionation that may result from sample degradation or issues of contamination arising from isotopic exchange with chemicals or artifacts from applied preservatives (Hobson *et al.* 1997).

Fixed tissues seemingly provide straightforward answers. However, as shown above, fixed tissues of migrants can have a temporal (and thus spatial) component that provides the researcher with an opportunity and a dilemma. The opportunity lies in the possibility of obtaining additional temporal or comparative information about the organism's movement because longitudinal isotope transects of the tissue will faithfully record this to some degree. The dilemma lies in obtaining the correct sample for addressing the research question at hand. Have we subsampled the correct part of the tissue to address the question? Which part of a slowly growing feather should we sample? Do we sample a small piece of feather from the top or bottom? Do we grind up a hair or whole feather? It is clear from the data in Figure 2.3 that random sampling near the top or bottom of a migrating bald eagle's flight feathers could produce very puzzling results. Knowledge of the biology of the organism under study and tissue growth patterns and timings are clearly essential.

Dynamic tissues, on the other hand, are those types of metabolically active tissues that are continuously "turning over" from an ongoing dietary source (Schoenheimer *et al.* 1939). Examples of these kinds of tissues are blood, muscle, liver, etc. While some of the stable isotope literature has focused on tissue turnover (Hobson and Clark 1992), very few controlled or laboratory studies have been conducted on the dynamic tissues of migrants as a means of gaining information about migratory connectivity. Most animal tissue turnover studies have focused on ^{13}C and ^{15}N (Trueman *et al.* 2005, Chamberlain *et al.* 2006), and fewer on other light isotopes (Hobson *et al.* 2004a). It is well established that living tissues in organisms can have vastly different isotopic turnover rates, ranging from days (liver, blood) to weeks (muscle) to lifetime (bone collagen). In theory, it should be possible to exploit the comparative multiple isotopic composition of dynamic tissues both temporally and spatially (cf. Tieszen *et al.* 1983, Suzuki *et al.* 2005, Cerling *et al.* 2006) in order to gain information about migration, but this remains an area of research that is just beginning to be explored. The development of various dynamic tissue isotopic turnover models based on laboratory experiments will be an essential prerequisite to the application in migration studies (Cerling *et al.* 2007).

Another overall consideration may be whether the fixed or dynamic tissue sample of interest can be collected nonlethally. For many threatened animals, hair, nail, and feather can easily be plucked or cut without fatally affecting the individual or its means of transport. For other species, euthanizing may be required, as in the case of small insects where entire wing sections must be taken. This may be a minor issue or could be a serious one in the case of rare or endangered species. Animal care guidelines and national regulatory guidelines must be observed when sampling.

Finally, the isotopic results obtained for both fixed and dynamic tissues will need to be related to on-the-ground spatial isotopic data (*e.g.*, isotopic base maps) in order to facilitate interpretations of geospatial movement. The assumptions and challenges in making these connections for migratory studies are fully discussed in Chapters 3–5.

In summary, the first step of capturing migrants for the purposes of applying stable isotopic assays is a major commitment and undertaking. However, equal and careful consideration must be given to the biology of the organism and to the type of tissue sample that is being collected from that individual for the purposes of stable isotope analyses. Consider carefully what kind of sample will be collected and what the results could mean. Stable isotopic assays do not deceive—it may be our understanding of the migrant's biology, biogeochemistry, or ecology that is lacking.

B. Cleaning, Storing, and Weighing

Once fixed or dynamic tissue samples have been taken from a migrant for stable isotope assays there follows the question of sample storage and preparation. The samples returned from the field may be dirty, matted, or greasy. Some degree of preparative cleaning may be required to remove external contamination. While the procedures for fixed tissues are fairly straightforward, there is ongoing debate concerning what to do with dynamic living tissues (*e.g.*, defatting, plasma separation).

For fixed tissues (hair, nail, claw), the sample cleaning procedure is uncomplicated. Ideally, the first goal would be to obtain the cleanest possible sample material directly from the animal. If there is dirt or other foreign adherents, samples may be washed in distilled water and air dried. Where there are natural oils on the fixed tissue (*e.g.*, oil on hair, feathers), samples should be further cleansed using a volatile solvent mixture. Here solvent cleansing is strongly recommended to remove surface oils because the carbon and hydrogen isotopic composition of oils and waxes can be markedly different (usually more negative) from pure keratinous tissue, and so could impart a disproportionate degree of isotopic variance. A recommended procedure for preparation of fixed tissues for light stable isotopes is outlined in Table 2.2. Note that natural oils on feathers typically do not contain S or N and so solvent cleaning may not be needed for these specific isotopic assays. For any trace elements (*e.g.*, Sr isotopes), sample cleaning and contamination of tissues is of major concern because contamination of the sample from dust and handling may be especially problematic (Font *et al.* 2007).

For dynamic tissues, the approach to cleaning procedures is less clear. A long-standing debate revolves around the fact that solvent or acid washes may change the isotopic composition of the bulk living tissue by selectively removing fatty acids or amino acids, and thereby changing the isotopic composition (CNS) of the bulk sample (Pinnegar and Polunin 1999, Sotiropoulos *et al.* 2004, Post *et al.* 2007). Opinions range from no cleaning, to solvent cleaning, to the use of C/N ratios to correct for fat content or empirical correction models. Currently there is no definitive one-size-fits-all cleaning procedure or correction approach that applies to all dynamic tissues. Nevertheless, the removal of lipids is highly preferable for fixed tissues given the potential variance in lipid $\delta^{13}C$ and δD values (Sessions *et al.* 1999). By removing fatty acids and lipids, we are assured that the isotopic analysis is conducted on only proteins.

Samples taken for stable isotopic analyses should be properly stored before and after preparative procedures in order to prevent degradation (and possible isotope fractionation). Many fixed tissue samples can be air or freeze-dried and stored in glass or plastic vials or paper envelopes at room temperature. For living tissues, samples can be stored frozen ($-40\,^\circ C$) to avoid decomposition, or freeze dried. If properly prepared and preserved, stable isotopic integrity over time will not be an issue.

TABLE 2.2	Recommended standard procedure for cleaning and weighing of "fixed" tissue for stable isotopic analysis

Working materials required: Analytical microbalance, cleaned tissue samples, clean culture tray(s), weighing utensils, methanol, Kimwipes, tray template, tape, marker, silver, or tin capsules

1. Obtain a clean 96 position plastic culture tray (Elisa Plate) and print out an Excel sample template.
2. Ensure feathers have been previously solvent cleaned [2:1 (v/v) chloroform/methanol 24 hour soak and 2× rinse] to remove surface oils. Air dry feathers in fume hood (>48 hours).
3. Cut off a small amount of feather vane (not rachis) material for analysis—always cut samples from the same location on different feather samples if possible for consistency (*e.g.*, sample near tip). Feather pieces can be cut using small stainless steel surgical scissors.
4. Clean weighing utensils using methanol and Kimwipes™ and allow to dry. Do not use acetone!
5. Make sure the microbalance is clean and calibrated. Ensure that the doors are closed when taring and weighing.
6. Tare a silver capsule,[a] handling only with tweezers, remove, and set on a clean metal surface. Use the smallest available capsule that will safely contain the sample (*e.g.*, 3.5 × 5.0 mm).
7. Using a spatula or tweezers, transfer a small amount of feather material into the capsule.
8. Reweigh, and continue adding or removing feather material until the target sample weight of 350 μg ± 10 μg is obtained.[b] This will take practice to get the feel of an appropriate amount. With practice, samples should take less than 5 min each to weigh out. Ensure the microbalance is accurate and stay within stated weight tolerance to avoid mass spectrometer mass-dependent source nonlinearity effects. Keratin or references must be weighed to the comparable elemental mass as the samples.
9. To seal the capsule, crimp the top of the capsule shut with a pair of straight edge tweezers and then fold tightly (as if folding down from the top of a paper bag). Then use the edge of the tweezers (use of two tweezers helps) to gently compact the capsule into a small, tight cube or ball. There should be no stray edges, loose sides, or feather material poking out. Flattened samples (rather than cube/ball-shaped) or capsules with stray or loose edges can jam our autosampler, cross-contaminate samples, and ruin an analysis.
10. Record final sample weight and sample name in spreadsheet. Place the sample capsule in the 96 position tray and record the weight on the tray template. Clean all utensils lightly with Kimwipes and methanol after completing each sample, air dry briefly. Secure the lid of the culture sample tray with rubber bands and masking tape and label the tray when done. Ensure samples cannot "jump" cells when the Elisa lid is properly closed (some brands of trays allow this).
11. Record sample name and weights (in milligrams) for each sample in the appropriate tray and its position (*e.g.*, tray 1, position A5). When completed, transfer this information to appropriate isotope laboratory sample submission form.
12. Use 3.5 × 5.0 mm silver or tin capsules designed for elemental isotope analysis. Suggested suppliers are Costech (1–800–524–7219) and Elemental Microanalysis (1–800–659–9885).

[a] Silver capsules must be used for δD and $\delta^{18}O$ analyses, tin capsules for $\delta^{13}C$, $\delta^{15}N$, and $\delta^{34}S$.
[b] Consult isotope analyst or laboratory being used for the specific mass to be used for each isotope. The example of feathers for C, H, N, O, and S assays is used, but it also applies to claw, hair, etc.

One exception noted above is the use of preservatives—it is not recommended to store samples in formalin as this has been shown to affect carbon and nitrogen isotope ratios (Hobson *et al.* 1997, Edwards *et al.* 2002). Formalin storage is a common problem for many museum specimens. If short-term solvent storage is required, instead use a 70% ethanol mixture.

C. Intra- and Intersample Heterogeneity

One question immediately facing the researcher in the field is the issue of which sample should be collected from the migrant organism for stable isotopic assays. For insects, the entire animal may need to be euthanized and taken. For nonlethal sampling of small and large birds or bats, there are a host of potential samples that can be considered. Fixed-tissue options include flight or contour feathers, hair, and nails. Dynamic-tissue options are typically blood, muscle, liver, and fats.

Questions have been raised regarding the issue of inter- and intrasample isotope heterogeneity (Wassenaar and Hobson 2006), where replicates of subsamples of feathers, hair, or other tissues may produce a range of C, H, N, O, and S isotopic values that could exceed the expected isotopic range that defines the organisms spatial distribution (Lott and Smith 2006). Studies have revealed that there are significant C, N, and H isotope differences among bulk tissue types (*e.g.*, blood, muscle, nails, feathers, hair) within a single individual (Tieszen *et al.* 1983, Mazerolle and Hobson 2005). Stable isotopic differences may be amplified at molecular levels if specific biochemical fractions (*e.g.*, lipids, amino acids) are further isolated (Teece and Vogel 2004). All of these intra- and intertissue isotopic differences arise from differential biochemical isotopic discriminations and from temporal changes in dietary sources that occur over the course of tissue biosynthesis (Phillips and Eldridge 2006). As noted above, each tissue type contains some shorter or longer term temporal record of diet (and likely each with differential diet-tissue isotopic fractionations) at a location or from many locations.

The researcher must be aware of both individual and population level *intersample isotopic heterogeneity*. There are two concerns. The first is the measurable C, H, N, O, and S isotopic differences that occur among the same tissues on the same animal. For example, due to inherent natural variance in isotope fractionations occurring during biochemical synthesis of tissues, we might expect there to be some minor isotopic variance, for example, among contour feathers grown by an individual at a single location. Second, we can expect intersample isotopic variance for the same feathers from a population of local birds that grew feathers at the same location. Intersample isotopic heterogeneity of both kinds are always greater than the instrumental analytical error for C, H, N, O, and S, and ideally would be much less than large-scale isotopic patterns in order to make geospatial interpretations (see Chapter 5).

Intrasample isotopic heterogeneity can be defined as the isotopic variance at the molecular (or microgram scale for our purposes here) level within a single discrete sample. This was illustrated above using δD and the eagle feather in Figure 2.3. Intrasample isotopic heterogeneity may be close to, but is always greater than, the instrumental analytical error for C, H, N, O, and S. A general rule-of-thumb is that large and slowly growing tissues will likely contain a greater degree of intrasample isotopic variance that will be amplified if the animal is moving across large spatial distances.

Several experiments with feathers and hair have shown that the overall level of δD variance associated with measurements strictly due to internal metabolic processes and laboratory methodology is of the order of $\pm 3‰$ (Bowen *et al.* 2005a, Wassenaar and Hobson 2006). Indeed, this is very close to the analytical error inherent in CF-IRMS measurements for δD ($\pm 2‰$). Comparable results are expected for other light isotopes. However, for most animals within- and among-tissue variance due to ecological considerations can greatly decrease the power of spatial resolution. In general, subsamples from selected tissues among individuals should be taken at the same location in order to represent, as best as possible, identical periods of growth.

Another issue that is commonly encountered in stable isotopic assays in migration studies is the problem of "outliers." For example, a researcher may submit a suite of feather samples for stable isotopic analysis from a local population of birds, and discover that all values are tightly clustered, except for a couple of extreme isotopic outliers. There are only two possible explanations, either the outliers are correct and must be explained or the data are faulty. The first thing the researcher should do is contact the laboratory to double check if any of the individual outlier analyses were faulty. A diligent analyst will welcome the query, even though the researcher may feel they are questioning the laboratory's integrity. Highly automated (100s of samples per day) IRMS systems are never foolproof and can have unexpected hardware glitches and mistakes in data handling. Fortunately, this is generally a rare occurrence. Nevertheless, the analyst will check the sample and standards QA/QC. If all seems to be OK, it may be worth requesting repeats for specific outliers (or by including subsample repeats in the first place). Once the outliers are confirmed to be correct, the researcher is left to ponder the ecological significance of the outliers (*e.g.*, recruitment of immigrants?).

In summary, the researcher must carefully consider which sample is the most likely to satisfy the requirements of answering, wholly or partially, the migration research question at hand. Even after careful selection and analysis of stable isotopes, a critical scrutiny of the results is warranted. This scrutiny will require knowledge of the biology of the species and may require some experimentation with the species under study, for example, to better quantify diet-tissue isotope fractionation. It is considered unwise to assume and extrapolate the experimental findings on intrasample isotopic heterogeneity among completely different species, as is aptly illustrated in Figure 2.3, although it has been noted that for many insectivorous passerines and waterfowl, the water-dietary-tissue fractionations for δD are remarkably uniform (Clark *et al.* 2006).

D. Sample Weighing

All samples submitted for stable isotope measurements require some form of sample processing, subsampling, and analytical weighing prior to their analysis. Many laboratories can provide this service and the client need not be overly concerned about the technical details. However, given the important cautions noted above regarding intrasample heterogeneity, clear and explicit instructions regarding exactly where subsamples need to be taken (*e.g.*, where on flight feathers) may have to be provided to the laboratory in order to avoid confusing results. Finally, a major cost savings and considerable improvement in sample turnaround time can usually be achieved if the researcher conducts the sample preparation and weighing. This should be done in full consultation with the stable isotope laboratory that will be employed.

Target weights for samples are typically grouped by the type of isotopic assay that is requested. For researchers that wish to conduct preparative weighing they will be requested by the laboratory to use a (costly) analytical microbalance. These balances are found in specialty analytical research laboratories and are capable of weighing samples to a readability of ± 0.001 mg. For each isotope, the mass of sample required will depend on the sensitivity of the instrument and the mass of the element in the sample.

Accurate weighing to the required laboratory target weight for the isotope of interest is absolutely essential. The main reason being that for isotopic assays by CF-IRMS, there is often a dependence of the δ result on sample mass because of differential gas pressures in the source of the IRMS, as demonstrated in Figure 2.4. This is commonly known as "source linearity," and is quantified in many laboratories for each gas species. For C + N, this mass dependency is typically low and forgiving. However, for isotopes like δD, this mass dependency can be significant, and easily up to 10‰ per 100 µg of sample in a positive or negative direction depending on the IRMS instrument. This means an accuracy of weighing to ± 10 µg is required to reduce this potential source of variance to below IRMS analytical error. Careless weighing (or poorly calibrated balances) is one of the first causes to consider for high variance in replicated results within and among laboratories. Hence, the researcher must enquire and adhere closely to the sample weight guidelines provided by the laboratory. If the selected laboratory has not established mass dependencies for their isotopic results, this is definitely the kind of QA/QC is worth asking about before doing the work. An example of the masses typically required for stable isotope assays are listed in Table 2.1.

III. GLOBAL-SPATIAL ISOTOPES

A. Stable-Hydrogen Isotopes

Stable-hydrogen isotope (δD) measurements are typically among the first to be considered when isotope assays for spatial analyses in migration studies are required (Hobson and Wassenaar 1997, Bowen *et al.* 2005b). Hydrogen isotopes have been shown to be especially robust in migration research

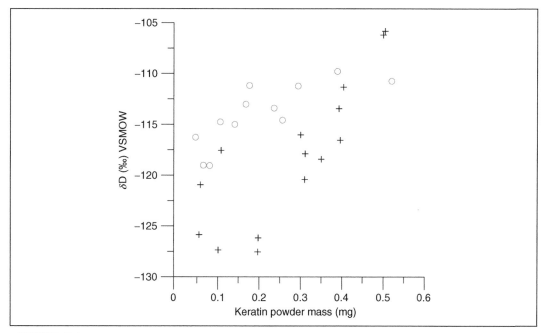

FIGURE 2.4 The dependence of δD results on mass from two different isotope ratio mass spectrometers at two different laboratories on isotopically homogenous keratin powder. Weighing error alone can result in δD changing at a rate of up to 10‰ per 100 µg of sample, well beyond acceptable error. From these data it is clear that sample and references must together be precisely and consistently weighted to a target weight ±10 µg.

applications (Wassenaar and Hobson 1998, Hobson 2002). There are several important and practical reasons why this is the case. First, there is extensive variation in δD (and $\delta^{18}O$) in nature between marine and terrestrial biomes, and across and among continents as a result of climatic and meteorological processes (Dansgaard 1964), and from equilibrium and kinetic isotope effects (Cormie *et al.* 1994, Hoefs 1997). These variations and patterns (which also apply to $\delta^{18}O$) are driven by well-known global meteorological processes and the equilibrium and kinetic isotope effects that occur spatially and that are strongly dependent on key environmental variables (temperature, elevation, rainout, prevailing source of moisture). In short, what this means is that for most continents, there are clear and predictable long-term geospatial isotopic patterns in the global H_2O cycle that are available to be translated into plants via primary productivity (Figure 2.2, see Chapter 4). Finally, the hydrogen isotope range in nature is large (~500‰), and compared to analytical error (less than ±2‰) gives by far the best signal to noise ratio of all of the light stable isotopes. In fact, δD has the sensitivity to resolve geospatial origins by a factor of three to five times better than $\delta^{18}O$ due to this factor alone.

B. The Problem of Hydrogen Isotope Exchange

Unfortunately, δD isotope analyses of organic substrates are more complicated than $\delta^{13}C$ and $\delta^{15}N$ because of the problem of uncontrolled hydrogen isotopic exchange between "labile" organic hydrogen in the sample matrix and ambient atmospheric water vapor (Schimmelmann 1991, Wassenaar and Hobson 2000b, Sessions and Hayes 2005). Although most of the hydrogen in fixed and dynamic tissues are not exchangeable and bound to carbon (C–H), a significant fraction of the total hydrogen, mainly in the form of –COOH, –NH$_2$, and –SH functional groups, readily exchanges hydrogen atoms with ambient water vapor (Schimmelmann *et al.* 1999). For proteins and keratins, the proportion of exchangeable hydrogen can be on the order of 15–25% or more of the total hydrogen (Wassenaar and Hobson 2000b, Bowen *et al.* 2005a), depending on temperature and how much of the exchangeable hydrogen is "exposed" by sample tissue grinding versus using whole sample pieces. Stable-hydrogen isotopic measurements of organic samples, however, can only be made on the total hydrogen in the sample. Therefore, unless there is some accounting and correcting for the exchangeable proportion of hydrogen, δD results will be incomparable among different laboratories where δD in local moisture is isotopically different and locally as the air changes seasonally.

C. The Comparative Equilibration Approach Facilitates δD Analyses

In the past, there was little consensus in the way isotope laboratories prepared, measured, and corrected for uncontrolled hydrogen isotopic exchange in organic samples. Nitration procedures aimed at replacing exchangeable hydrogen are useful for few substrates (cellulose, chitin) and require extensive processing of individual samples (Schimmelmann and DeNiro 1986). Nitration was not applicable to keratins and other dynamic and fixed tissues. Steam equilibration with H_2O of known isotopic composition was successful when dual-inlet assays were prevalent, but this method was extremely slow (10 samples per day) and expensive due to high degree of labor by technicians. Several years ago the "comparative equilibration" approach using CF-IRMS was proposed for δD analyses of organic materials for migration research (Wassenaar and Hobson 2003). The key benefits of this approach are comparable results among laboratories coupled with ease of application and rapid and automated sample throughput (200 samples per day), as increasingly demanded by the large numbers of samples typically generated in migration studies.

The comparative equilibration approach takes advantage of the fact that hydrogen isotope exchange between ambient laboratory air moisture and the exchangeable hydrogen in keratins is fast, and will reach full equilibrium by ~96 hours at room temperatures, even when samples are fully weighed and prepared in their silver capsules (Wassenaar and Hobson 2000a, Bowen *et al.* 2005a). The comparative aspect means the inclusion of precalibrated keratin working standards (see below) along with all unknown tissue samples. As the ambient moisture changes its δD, both keratinous samples and references will equilibrate in identical fashion. These "comparatively equilibrated" keratin standards and unknowns are then isolated from the atmosphere (using a zero-blank autosampler) and analyzed together in a single analysis session. This approach follows the principle of identical treatment for stable isotope analyses, whereby samples and working standards are not only identical in their chemical composition, but go through exactly the same preparation and analysis steps (Werner and Brand 2001).

In most laboratories, stable-hydrogen isotope measurements on organic tissues are currently performed on H_2 derived from high-temperature flash pyrolysis and by CF-IRMS (Figure 2.1B). Pure H_2 is used as the sample analysis gas and the isotopic reference gas. A high-temperature EA and autosampler is used to pyrolyze samples to a single pulse of H_2 gas (and N_2 and CO gas). The pyrolysis column consists of a ceramic tube partially filled with glassy carbon chips held at 1250–1350 °C, followed by a molecular sieve GC column at 80–100 °C. The GC column is used to resolve the sample H_2 from N_2 and CO. All δD results are reported in units of per mil (‰) relative to the VSMOW-SLAP standard scale using previously calibrated keratin or organic working standards.

D. The Problem of Organic Reference Materials for δD

The success of the comparative equilibration approach for δD assays relies on the long-term availability and widespread distribution of previously (steam) calibrated keratin and organic tissue working standards. The reason is that intercomparability of results among laboratories and studies is critical for ensuring quality and accuracy of results. While a number of primary and certified organic standards currently exist (IAEA CH-7, etc), these certified primary reference materials do not contain exchangeable hydrogen, and so are not suitable reference materials for calibrating samples that do have exchangeable hydrogen.

For this reason, three keratin laboratory working standards were previously prepared and distributed among a number of laboratories. These keratins were composed of cryogenically ground and isotopically homogenized chicken feather (CFS), cow hoof (CHS), and bowhead whale baleen (BWB-II) that were calibrated to account for exchangeable hydrogen using an off-line steam equilibration and dual inlet assays as outlined by Wassenaar and Hobson (2003). CFS was obtained from a single batch of chicken feathers from a poultry processing operation located in Wynyard, Saskatchewan, Canada. Approximately 2 kg of feathers were obtained and processed. The CHS (~0.5 kg) was obtained by cutting hoof from a single cow carcass at an abattoir in Saskatoon, Canada. The BWB-II was a powdered whale baleen obtained from Professor Don Schell (retired) from the University of Alaska. All of these keratins were solvent cleansed (2:1 chloroform:methanol solution), air-dried, and cryogenically homogenized to large quantities (>0.5 kg). Sieving to the <100 µm fraction and further blending was required to ensure isotopic homogeneity at the 100 µg level. Sufficient standards were prepared to ensure years of use in a single laboratory. The following results for δD are accepted: CFS $= -147 \pm 5$‰ (VSMOW), CHS $= -187 \pm 2$ (VSMOW), and BWB-II $= -108 \pm 4$‰ (VSMOW). It should be emphasized that these keratins are not primary isotopic reference materials for organic δD analyses. They are a rapidly dwindling supply of unofficial laboratory working standards to be used for comparative equilibration.

Concerted efforts are urgently needed to produce isotopic working standards for δD of keratinous and other organic matrices of ecological interest that are not only suitable for comparative equilibration, but will be available for the long-term need of migration and ecological research. Currently, the range of encountered δD values for keratins in nature greatly exceeds the CHS-CFS-BWB calibration range. This will require a global search and preparation of at least three keratin and organic working standards that meet the following criteria: (1) sufficient quantities to meet the demand of many stable isotope laboratories internationally for at least 10 years, (2) certified stable isotopic homogeneity for δD at the <100 µg level, and (3) a δD isotopic range of at least 200‰, spanning from about +50‰ to −200‰. At the time of this writing, no certified or internationally accepted organic standards currently exist for δD for matrices that have exchangeable hydrogen. A number of isotope laboratories have attempted to widen availability by grounding the results of bulk commercial keratin powders and or other types of local or broadly available keratinous standards to the BWB-CFS-CHS scale (*e.g.*, Table 2.3), or by producing internal working keratin standards based on their own calibration curves. However, this problem has largely been addressed on an ad hoc basis for the short-term needs of a few laboratories or researchers for specific projects. This problem of standardization for organic matrices

TABLE 2.3 Laboratory intercomparison of δD results for powdered keratin samples run over many months using the comparative equilibration approach

	Laboratory 1 mean	Laboratory 1 SD	*n*	Laboratory 2 mean	Laboratory 2 SD	*n*
Moose hair	−163.5	2.1	84	−164.7	2.4	54
Vole hair	−106.2	2.3	85	−105.1	2.6	21
Human hair (IAEA-085)	−70	2.4	84	−68.7	2.6	51
Keratin (Spectrum)	−117.2	1.9	103	−116.1	2.7	106

The results reveal excellent repeatability and comparability of results among latitudinally separated laboratories. All samples were treated as unknowns and normalized using CHS, CFS, and BWB keratin laboratory standards. (Courtesy T. Jardine and R. Doucette, unpublished data.)

with exchangeable hydrogen will become pressing in the near future as interest in using δD in migration and forensic research diversifies and increases, and the demand by new laboratories increases and isotopic approaches move out of the realm of academics to become a mainstream tool for decision makers. Further, directly comparable results will be required as data from numerous migration studies accumulate over time and the potential for meta-analysis becomes possible. The problems of reference materials for organic-H samples currently remain a formidable and outstanding challenge.

A number of laboratory intercomparison tests on discrete samples and homogenized keratin powders have been conducted using the comparative equilibration approach. The data reveal that laboratories can consistently achieve comparable results on homogenized standards to within an acceptable error of $\pm$2‰ for δD. Another example of excellent agreement obtained by using comparative equilibration is shown in Figure 2.5. Here independent laboratory intercomparison of δD assays was made on single feathers from 18 individual songbirds using the comparative equilibration procedure approach. All laboratories used the keratin laboratory standards described below. However, no attempt was made here to homogenize the feathers (*e.g.*, subsamples were cut from the same feather at each laboratory with no consideration of location on the feather), and because the samples were not examined for isotopic heterogeneity the average δD range per sample was 7‰, which is remarkably good. An excellent example of a long-term test from two laboratories using the comparative equilibration approach for δD is shown in Table 2.3.

E. Stable Oxygen Isotopes

Stable-oxygen isotopes (δ^{18}O) may also be considered for global-spatial analyses in migration studies for precisely the same reasons as δD noted above (Hobson *et al.* 2004b, Bowen *et al.* 2005b). A key advantage to δ^{18}O is that organic materials like keratins have no exchangeable oxygen, circumventing the problem and need for comparative equilibration approaches. The main disadvantage, however, is that the δ^{18}O range for tissues in nature is relatively small ($\sim$15‰ range) and the analytical error by CF-RIMS methods currently remains comparatively high ($\pm$0.5‰), resulting in a lowered precision in resolving geospatial information. However, there is still a paucity of δ^{18}O data on migrant tissues, and the use of oxygen isotopes, especially in concert with hydrogen isotopes, remains promising, but largely unexplored.

In most modern laboratories, stable oxygen isotope measurements on organic tissues are performed on CO derived from high-temperature pyrolysis and by CF-IRMS (Figure 2.1B). Pure CO is used as the

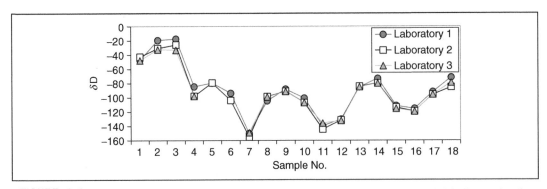

FIGURE 2.5 Independent laboratory intercomparison of δD assays on 18 individual songbird feathers using the comparative equilibration procedure approach. No attempt was made to homogenize the feathers (*e.g.*, subsamples were cut from the same feather at each laboratory with no consideration of location on the feather), and because the samples were not screened for heterogeneity, the average δD range per sample was 7‰ (additional data from T. Jardine and R. Doucett).

sample analysis gas and the isotopic reference gas. A high-temperature EA and autosampler is used to automatically pyrolyze samples to a single pulse of CO gas (and N_2 and H_2 gas). The pyrolysis column consists of a ceramic tube and glassy carbon tube insert, filled to the hot zone with glassy carbon chips held at >1350 °C, followed by a molecular sieve GC column at 40–60 °C. The GC column is used to resolve the sample CO from H_2 and N_2. All $\delta^{18}O$ results are reported in units of per mil (‰) relative to the VSMOW-SLAP standard scale using two newly available primary and certified organic reference materials (IAEA Benzoic Acid Standards). These primary standards have $\delta^{18}O$ values of +23.1‰ VSMOW (IAEA-601) and +71.4‰ VSMOW (IAEA-604). It should be noted that the $\delta^{18}O$ of biological tissues from migrants range between +10 and +20‰, below this primary reference calibration range, and while this calibration is highly linear, the development of keratin references for $\delta^{18}O$ spanning the natural range is encouraged.

The reader may have noticed that the CF-IRMS analytical methods for H_2 and CO described above are virtually identical, and because H_2 is produced during the same thermochemical reduction reaction as for CO, preparative systems have been tweaked in order to obtain $\delta^{18}O$ and δD on a single tissue sample. While this appears to be an attractive option, the primary challenge is lowered sample throughput and a more complicated analytical setup involving multiple reference gases and IRMS peak jumping. The analysis time for dedicated δD assays is rapid at about 3 min per sample (20 samples/hour), but when combined to obtain both $\delta^{18}O$ and δD, the analysis rate decreases to 8–12 min per sample. Given that $\delta^{18}O$ is expected to provide correlative results to δD with lowered precision, this approach may be questioned as to its practical utility. However, it should be noted that additional information may be obtained from $\delta^{18}O$ because there are more oxygen sources (air O_2, H_2O, dietary O) and sinks (H_2O, CO_2) in biological dietary systems compared to hydrogen. This additional complexity (and associated isotope fractionations) may prove to be of use in the future with more research.

IV. LOCAL-SPATIAL ISOTOPES

A. Stable Carbon and Nitrogen Isotopes

Stable-carbon and -nitrogen isotope assays are considered local-spatial analyses, but may be very useful in further delineating migratory populations or indicating type of habitat (see Chapter 1). There may also be larger-scale spatial patterns (Chapter 4). The analysis of carbon and nitrogen isotopes are almost universally considered routine among stable isotopes laboratories.

The analyses are composed of coupled $\delta^{13}C$ and $\delta^{15}N$ isotope measurements on the same organic tissue sample performed on CO_2 and N_2 derived from flash combustion and by CF-IRMS (Figure 2.1A). Pure CO_2 and N_2 are used as the sample analysis gas and the isotopic reference gas. A standard EA and autosampler is used to quantitatively combust samples to a pulse of CO_2 and N_2 gas (combustion H_2O is scrubbed out with a trap). The oxidation column consists of a quartz tube partially filled to the hot zone with chromium oxide held at 1050 °C, followed by reduction column filled with copper (to reduce NO_x to N_2) held at 600–800 °C, and then a packed GC column at 35–50 °C. The GC column with a thermal conductivity detector is used to quantify and resolve CO_2 from N_2. Isotope peak jumping is used on the IRMS to switch between nitrogen and carbon isotopes. All $\delta^{13}C$ results are reported in units of per mil (‰) relative to the PDB using newly available primary certified organic reference materials (L-glutamic acids). These standards have $\delta^{13}C$ values of +37.6‰ VPDB (USGS-41) and −26.4‰ PDB (USGS-40) and $\delta^{15}N$ values of +47.6‰ (AIR) and −4.5‰ (AIR), respectively. The standards have similar C/N ratios as proteinaceous tissues, and fully span the isotopic range encountered for migrant tissues in nature.

B. Stable Sulfur Isotopes

Stable sulfur isotope analyses for use in migration research are still relatively rare, and can also be grouped into the local-spatial analyses category. The range of $\delta^{34}S$ in nature is very large (spanning > 150‰), although it should be noted that the $\delta^{34}S$ of the seawater sulfate pool in the oceans is essentially invariant (+21‰ VCDT), and that terrestrial systems tend to have negative $\delta^{34}S$ compositions, making S isotopes useful for detecting or distinguishing between marine and terrestrial dietary sources.

For the tissues of interest in migratory organisms, sulfur is primarily in the form of S-bearing amino acids (*e.g.*, cysteine). As mentioned previously, the best precision $\delta^{34}S$ analyses are still made using conventional dual-inlet IRMS. This requires a lengthy and costly preparative process that involves oxidative and quantitative conversion of total S in the tissue sample to an appropriate analyte (*e.g.*, $BaSO_4$, Ag_2S). These matrices are then converted to SO_2 or SF_6 gas for analyses on a gas source dual-inlet IRMS (Mayer and Krouse 2004).

More recently, CF-IRMS methods have been developed, with the advantage of lower cost and higher sample throughput. Similarly to C + N, tissue samples that have been converted to $BaSO_4$ or Ag_2S are combusted in an EA, and the SO_2 produced is chemically purified and separated from CO_2 and N_2 using a GC column (Mayer and Krouse 2004). This analysis is made on $BaSO_4$ or Ag_2S and so still requires preparative sample conversions. The precision for this method is about ±0.2‰, and requires about 15 min per sample (<100 µg S).

There is also immense interest in the direct EA combustion of tissue and keratin samples to SO_2 without conversion to $BaSO_4$ or Ag_2S. This approach is achievable but is complicated by high C:S ratios of most samples that generate comparatively large amounts of CO_2 and H_2O and subsequent separation of the undesired combustion gases by GC. However, recent innovations and adaptations have enabled concurrent C + N + S assays (Fry 2007). The general consensus is that direct combustion and analysis is only feasible where samples have >0.1 wt.% S (requiring 2–5 mg of keratin), and the resulting $\delta^{34}S$ precisions will be on the order of ±0.5‰, although this may be sufficient for some studies. Analytical vigilance is required because the $\delta^{18}O$ of the SO_2 produced changes as reagents are depleted, requiring dynamic oxygen corrections to obtain correct $\delta^{34}S$ values (Fry *et al.* 2002).

A further complication is that keratin working standards for organic-S are currently nonexistent, and large discrepancies occur when comparing inorganic standards to organic samples. Ideally, a set of keratin (*e.g.*, cysteine) or organic S standards would need to be developed that had been previously converted to $BaSO_4$ or Ag_2S and measured by highest precision dual-inlet analyses, and that span the natural $\delta^{34}S$ range. However, given the large sample size requirements (2–5 mg) and the need to include many references within a CF-IRMS autorun, it is likely that calibration and organic reference material development will remain within the realm of a few specialized individual laboratories.

C. Stable Isotopes of Trace Elements: $^{87}Sr/^{86}Sr$

Another type of local-spatial stable isotope analysis that has been used in animal migration research is the isotopes of the trace element strontium ($^{87}Sr/^{86}Sr$) (Chamberlain *et al.* 1997). One of the key advantages of "heavy isotopes" compared to the light stable isotopes of the previous sections is that there is usually little or no isotopic fractionation from geologic sources through the food web and into tissues (Blum *et al.* 2001). Hence, Sr isotopes among all species in local food webs are expected to show fidelity to the $^{87}Sr/^{86}Sr$ ratios of the underlying bedrock or soil. Thus, $^{87}Sr/^{86}Sr$ variations in and among landscapes and continents may be distinctive, but these can be highly variable at small scales in areas of complex geology and are not *a priori* suitable for continuous interpolation (Chapter 4). As with S, the $^{87}Sr/^{86}Sr$ distinctions between terrestrial and marine environments are very clear (Kennedy *et al.* 2005).

Hence, similarly to the other local-spatial isotopes, Sr isotopes are best used in conjunction with δD or $\delta^{18}O$ or other light isotopes in a multiisotope approach.

One important difference between the light isotopes and Sr isotopes in fixed tissues like feathers is that Sr occurs as a trace element at exceedingly low concentrations (*e.g.*, <20 μg Sr/g of feather) (Font *et al.* 2007). The only exception is calcium-bearing tissues that contain much higher Sr concentrations (*e.g.*, bones, teeth, otoliths) (Blum *et al.* 2001). For noncalcium-bearing tissues, this means the potential for extraneous contamination (*e.g.*, entrained dust, handling, background) is extremely high and must be quantified and requires rigorous cleaning and QA/QC procedures (Font *et al.* 2007). There is no standardized agreement on how this should be done.

Further, trace Sr concentrations means that up to 25 mg of sample may be required. All samples will require prescreening for total Sr content by inductively coupled-plasma mass spectrometry before preparative procedures for isolating Sr are begun. Further, extensive wet chemical or microwave digestions and selective ion chromatography are required to isolate Sr for isotopic analysis. Difficulties arise with organic samples in being able to fully extract all available Sr. Sr isotopic ratios are determined using thermal ionization mass spectrometry and using the NBS-987 standard. Similar to S, no organic-Sr isotope standards exist.

The issue of intrasample variability has not been rigorously tested for Sr isotopes, although differences in concentrations and isotope ratios between the rachis and vane of individual feathers have been reported (Font *et al.* 2007), although these differences were comparatively smaller than potential geospatial differences. From the preceding section, this could be a major issue for slowly growing tissues of birds or animals that are moving among areas having variable $^{87}Sr/^{86}Sr$, especially given the large sample requirements. For example, when we consider that feathers of small birds only weigh 10–20 mg, or hair strands 1 mg or less, sample tissue pooling may be required. The example of the eagle feather heterogeneity and the implications for confounding Sr isotope interpretations regarding spatial interpretations are critical to consider.

Finally, Sr isotope analyses are costly in comparison to the light isotopes. Current costs do not include the costs of prescreening for Sr concentration nor the labor involved in clean-laboratory wet chemical digestions. This, and given the large sample size requirements, it is likely that Sr isotope assays may remain a specialized assay for projects where the value of the anticipated outcomes exceeds the analytical cost considerations.

V. CONCLUSIONS

In this chapter, some of the fundamental and practical aspects of stable isotope analyses for use in animal migration research have been outlined for those interested in applying these isotopic tracers to research questions. The researcher is strongly encouraged to be discriminating and critical in the application of stable isotopes, and to carefully consider all angles of sampling and selection of isotopes to answer the research question at hand. Only when utmost confidence in the stable isotope analyses is fully assured, then the researcher can move on to the task of making the kinds of spatial interpretations outlined in the subsequent chapters.

VI. ACKNOWLEDGMENTS

Special thanks to Tim Jardine (University of New Brunswick) and Rick Doucett (Northern Arizona University) for contributing data to Figure 2.4, 2.5, and Table 2.3. Feathers for Figure 2.5 were provided by Tony Diamond (University of New Brunswick). Thanks to Gabe Bowen, John Matthews, Tim Jardine, and Keith Hobson for helpful comments and suggestions.

VII. REFERENCES

Blum, J. D., E. H. Taliaferro, and R. T. Holmes. 2001. Determining the sources of calcium for migratory songbirds using stable strontium isotopes. *Oecologia* **126**(4):569–574.

Bowen, G. J., L. Chesson, K. Nielson, T. E. Cerling, and J. R. Ehleringer. 2005a. Treatment methods for the determination of δ^2H and $\delta^{18}O$ of hair keratin by continuous-flow isotope-ratio mass spectrometry. *Rapid Communications in Mass Spectrometry* **19**(17):2371–2378.

Bowen, G. J., L. I. Wassenaar, and K. A. Hobson. 2005b. Global application of stable hydrogen and oxygen isotopes to wildlife forensics. *Oecologia* **143**(3):337–348.

Cerling, T. E., G. Wittemyer, H. B. Rasmussen, F. Vollrath, F. Cerling, T. J. Robinson, and I. Douglas-Hamilton. 2006. Stable isotopes in elephant hair document migration patterns and diet changes. *Proceedings of the National Academy of Sciences of the United States of America* **103**(2):371–373.

Cerling, T. E., L. K. Ayliffe, M. D. Dearing, J. R. Ehleringer, B. H. Passey, D. W. Podlesak, A. M. Torregrossa, and A. G. West. 2007. Determining biological tissue turnover using stable isotopes: The reaction progress variable. *Oecologia* **151**(2):175–189.

Chamberlain, C. P., J. D. Blum, R. T. Holmes, X. H. Feng, T. W. Sherry, and G. R. Graves. 1997. The use of isotope tracers for identifying populations of migratory birds. *Oecologia* **109**(1):132–141.

Chamberlain, P. M., I. D. Bull, H. I. J. Black, P. Ineson, and R. P. Evershed. 2006. The effect of diet on isotopic turnover in *Collembola* examined using the stable carbon isotopic compositions of lipids. *Soil Biology & Biochemistry* **38**(5):1146–1157.

Clark, I. D., and P. Fritz. 1997. *Environmental Isotopes in Hydrogeology*. Lewis Publishers, New York.

Clark, R. G., K. A. Hobson, and L. I. Wassenaar. 2006. Geographic variation in the isotopic (δD, $\delta^{13}C$, $\delta^{15}N$, $\delta^{34}S$) composition of feathers and claws from lesser scaup and northern pintail: Implications for studies of migratory connectivity. *Canadian Journal of Zoology-Revue Canadienne De Zoologie* **84**(10):1395–1401.

Cormie, A. B., H. P. Schwarcz, and J. Gray. 1994. Determination of the hydrogen isotopic composition of bone collagen and correction for hydrogen exchange. *Geochimica et Cosmochimica Acta* **58**:365–375.

Criss, R. E. 1999. *Principles of Stable Isotope Distribution*. Oxford University Press, New York.

Dansgaard, W. 1964. Stable isotopes in precipitation. *Tellus* **5**:436–468.

de Groot, P. A. 2004. *Handbook of Stable Isotope Analytical Techniques*. 1st edn. Elsevier, Amsterdam.

Duxbury, J. M., G. L. Holroyd, and K. Muehlenbachs. 2003. Changes in hydrogen isotope ratios in sequential plumage stages: An implication for the creation of isotope-base maps for tracking migratory birds. *Isotopes in Environmental and Health Studies* **39**(3):179–189.

Edwards, M. S., T. F. Turner, and Z. D. Sharp. 2002. Short- and long-term effects of fixation and preservation on stable isotope values ($\delta^{13}C$, $\delta^{15}N$, $\delta^{34}S$) of fluid-preserved museum specimens. *Copeia* **4**:1106–1112.

Font, L., G. M. Nowell, D. G. Pearson, C. J. Ottley, and S. G. Willis. 2007. Sr isotope analysis of bird feathers by TIMS: A tool to trace bird migration paths and breeding sites. *Journal of Analytical Atomic Spectrometry* **22**(5):513–522.

Fry, B. 2006. *Stable Isotope Ecology*. Springer, New York.

Fry, B. 2007. Coupled N, C and S stable isotope measurements using a dual-column gas chromatography system. *Rapid Communications in Mass Spectrometry* **21**(5):750–756.

Fry, B., W. Brand, F. J. Mersch, K. Tholke, and R. Garritt. 1992. Automated-analysis system for coupled delta-c-13 and delta-n-15 measurements. *Analytical Chemistry* **64**(3):288–291.

Fry, B., S. R. Silva, C. Kendall, and R. K. Anderson. 2002. Oxygen isotope corrections for online $\delta^{34}S$ analysis. *Rapid Communications in Mass Spectrometry* **16**(9):854–858.

Groning, M. 2004. International stable isotope reference materials. Pages 874–906. , P. D. Groot (Ed.) *Handbook of Stable Isotope Analytical Techniques*, Volume 1. Elsevier, Amsterdam.

Hebert, C. E., and L. I. Wassenaar. 2001. Stable nitrogen isotopes in waterfowl feathers reflect agricultural land use in western Canada. *Environmental Science & Technology* 35(17):3482–3487.

Hebert, C. E., and L. I. Wassenaar. 2005. Feather stable isotopes in western North American waterfowl: Spatial patterns, underlying factors, and management applications. *Wildlife Society Bulletin* 33 (1):92–102.

Hobson, K. A. 1999. Tracing origins and migration of wildlife using stable isotopes: A review. *Oecologia* 120:314–326.

Hobson, K. A. 2002. Incredible journeys. *Science* 295(5557):981.

Hobson, K. A., and D. M. Schell. 1998. Stable carbon and nitrogen isotope patterns in baleen from eastern Arctic bowhead whales (*Balaena mysticetus*). *Canadian Journal of Fisheries and Aquatic Sciences* 55(12):2601–2607.

Hobson, K. A., and L. I. Wassenaar. 1997. Linking breeding and wintering grounds of neotropical migrant songbirds using stable hydrogen isotopic analysis of feathers. *Oecologia* 109:142–148.

Hobson, K. A., and R. G. Clark. 1992. Assessing avian diets using stable isotopes 2. Factors influencing diet-tissue fractionation. *Condor* 94(1):189–197.

Hobson, K. A., H. L. Gibbs, and M. L. Gloutney. 1997. Preservation of blood and tissue samples for stable-carbon and stable-nitrogen isotope analysis. *Canadian Journal of Zoology-Revue Canadienne De Zoologie* 75(10):1720–1723.

Hobson, K. A., L. Atwell, L. I. Wassenaar, and T. Yerkes. 2004a. Estimating endogenous nutrient allocations to reproduction in Redhead Ducks: A dual isotope approach using δD and $\delta^{13}C$ measurements of female and egg tissues. *Functional Ecology* 18(5):737–745.

Hobson, K. A., G. J. Bowen, L. I. Wassenaar, Y. Ferrand, and H. Lormee. 2004b. Using stable hydrogen and oxygen isotope measurements of feathers to infer geographical origins of migrating European birds. *Oecologia* 141(3):477–488.

Hoefs, J. 1997. Stable Isotope Geochemistry. *completely rev., pdated, and enl. edn* 4th Springer, Berlin, New York.

Hoefs, J. 2004. *Stable Isotope Geochemistry.* 5th edn Springer, Berlin, New York.

Kendall, C., and J. J. McDonnell. 1998. *Isotope Tracers in Catchment Hydrology.* Elsevier, New York.

Kennedy, B. P., C. P. Chamberlain, J. D. Blum, K. H. Nislow, and C. L. Folt. 2005. Comparing naturally occurring stable isotopes of nitrogen, carbon, and strontium as markers for the rearing locations of Atlantic salmon (*Salmo salar*). *Canadian Journal of Fisheries and Aquatic Sciences* 62(1):48–57.

Lajtha, K., and R. Michener. 2007. *Stable Isotopes in Ecology and Environmental Science.* Blackwell Scientific, New York.

Lott, C. A., and J. P. Smith. 2006. A geographic-information-system approach to estimating the origin of migratory raptors in North America using stable hydrogen isotope ratios in feathers. *Auk* 123 (3):822–835.

Matthews, D. E., and J. M. Hayes. 1978. Isotope-ratio-monitoring gas chromatography-mass spectrometry. *Analytical Chemistry* 50(11):1465–1473.

Mayer, B., and H. R. Krouse. 2004. Procedures for sulfur isotope abundance studies. Pages 538–596. , P. A. De Groot (Ed.) *Handbook of Stable Isotope Analytical Techniques*, Volume 1. Elsevier, Amsterdam.

Mazerolle, D. F., and K. A. Hobson. 2005. Estimating origins of short-distance migrant songbirds in North America: Contrasting inferences from hydrogen isotope measurements of feathers, claws, and blood. *Condor* 107(2):280–288.

McKinney, C. R., J. M. McCrea, S. Epstein, H. A. Allen, and H. C. Urey. 1950. Improvements in mass spectrometers for the measurement of small differences in isotope abundance ratios. *Review of Scientific Instruments* 21(8):724–730.

O'Brien, D., and M. J. Wooller. 2007. Tracking human travel using stable oxygen and hydrogen isotope analyses of hair and urine. *Rapid Communications in Mass Spectrometry* 21:2422–2430.

Phillips, D. L., and P. M. Eldridge. 2006. Estimating the timing of diet shifts using stable isotopes. *Oecologia* 147(2):195–203.

Pinnegar, J. K., and N. V. C. Polunin. 1999. Differential fractionation of delta $\delta^{13}C$ and $\delta^{15}N$ among fish tissues: Implications for the study of trophic interactions. *Functional Ecology* 13(2):225–231.

Post, D. M., C. A. Layman, D. A. Arrington, G. Takimoto, J. Quattrochi, and C. G. Montana. 2007. Getting to the fat of the matter: Models, methods and assumptions for dealing with lipids in stable isotope analyses. *Oecologia* 152(1):179–189.

Schimmelmann, A. 1991. Determination of the concentration and stable isotopic composition of nonexchangeable hydrogen in organic matter. *Analytical Chemistry* 63:2456–2459.

Schimmelmann, A., and M. J. DeNiro. 1986. Stable isotopic studies on chitin. III. The D/H and $^{18}O/^{16}O$ ratios in arthropod chitin. *Geochimica et Cosmochimica Acta* 50:1485–1496.

Schimmelmann, A., M. D. Lewan, and R. P. Wintsch. 1999. D/H isotope ratios of kerogen, bitumen, oil and water in hydrous pyrolysis of source rocks containing kerogen types-I, -II, -IIS, and -III. *Geochimica et Cosmochimica Acta* 63(22):3751–3766.

Schoenheimer, R., S. Ratner, and D. Rittenberg. 1939. Studies in protein metabolism X. The metabolic activity of body proteins investigated with l (−)-leucine containing two isotopes. *Journal of Biological Chemistry* 130(2):703–732.

Sessions, A. L., and J. M. Hayes. 2005. Calculation of hydrogen isotopic fractionations in biogeochemical systems. *Geochimica Et Cosmochimica Acta* 69(3):593–597.

Sessions, A. L., T. W. Burgoyne, A. Schimmelmann, and J. M. Hayes. 1999. Fractionation of hydrogen isotopes in lipid biosynthesis. *Organic Geochemistry* 30(9):1193–1200.

Sharp, Z. 2007. *Stable Isotope Geochemistry*. Prentice-Hall, New Jersey.

Sotiropoulos, M. A., W. M. Tonn, and L. I. Wassenaar. 2004. Effects of lipid extraction on stable carbon and nitrogen isotope analyses of fish tissues: Potential consequences for food web studies. *Ecology of Freshwater Fish* 13(3):155–160.

Suzuki, K. W., A. Kasai, K. Nakayama, and M. Tanaka. 2005. Differential isotopic enrichment and half-life among tissues in Japanese temperate bass (Lateolabrax japonicus) juveniles: Implications for analyzing migration. *Canadian Journal of Fisheries and Aquatic Sciences* 62(3):671–678.

Teece, M., and M. Vogel. 2004. Preparation of ecological and biological samples for isotope analysis. Pages 177–202. P. A. de Groot (Ed.) *Handbook of Stable Isotope Techniques*. Elsevier, Amsterdam.

Tieszen, L. L., T. W. Boutton, K. G. Tesdahl, and N. A. Slade. 1983. Fractionation and turnover of stable carbon isotopes in animal tissues: Implications for $\delta^{13}C$ analysis of diet. *Oecologia* 57:32–37.

Trueman, C. N., R. A. R. McGill, and P. H. Guyard. 2005. The effect of growth rate on tissue-diet isotopic spacing in rapidly growing animals. An experimental study with Atlantic salmon (Salmo salar). *Rapid Communications in Mass Spectrometry* 19(22):3239–3247.

Wassenaar, L. I., and K. A. Hobson. 1998. Natal origins of migratory monarch butterflies at wintering colonies in Mexico: New isotopic evidence. *Proceedings of the National Academy of Sciences of the United States of America* 95(26):15436–15439.

Wassenaar, L. I., and K. A. Hobson. 2000a. Improved method for determining the stable-hydrogen isotopic composition (δD) of organic materials of environmental interest. *Environmental Science and Technology* 34:2354–2360.

Wassenaar, L. I., and K. A. Hobson. 2000b. Improved method for determining the stable-hydrogen isotopic composition (delta D) of complex organic materials of environmental interest. *Environmental Science and Technology* 34(11):2354–2360.

Wassenaar, L. I., and K. A. Hobson. 2003. Comparative equilibration and online technique for determination of non-exchangeable hydrogen of keratins for use in animal migration studies. *Isotopes in Environmental and Health Studies* 39(3):211–217.

Wassenaar, L. I., and K. A. Hobson. 2006. Stable-hydrogen isotope heterogeneity in keratinous materials: Mass spectrometry and migratory wildlife tissue subsampling strategies. *Rapid Communications in Mass Spectrometry* 20(16):2505–2510.

Werner, R. A., and W. A. Brand. 2001. Referencing strategies and techniques in stable isotope ratio analysis. *Rapid Communications in Mass Spectrometry* 15(7):501–519.

CHAPTER 3

Applying Isotopic Methods to Tracking Animal Movements

Keith A. Hobson

Environment Canada

Contents

I. INTRODUCTION

Stable isotope methods have proven to be a boon to ecologists in the last decades as witnessed by the veritable explosion of papers dealing with a diversity of taxa ranging from invertebrates to polar bears. This follows on from a well-established tradition of development and adoption of stable isotope techniques seen in other disciplines such as the earth sciences, plant physiology, and anthropology. The vast majority of the more recent contributions have focused on dietary reconstructions or trophic interactions because stable isotope measurements can provide information on sources of nutrients to food webs and trophic level according to the perhaps overused maxim that "you are what you eat plus a few parts per mil." More recently, attention has turned to applying stable isotope measurements as

Tracking Animal Migration with Stable Isotopes
K. A. Hobson and L. I. Wassenaar (Editors)
ISSN 1936-7961, DOI: 10.1016/S1936-7961(07)00003-6

forensic tracers of origin for migratory or dispersing individual organisms. This development relies on three basic isotopic principles:

1. Consumer (animals including humans) stable isotope values reflect those of the food web they are in equilibrium with. Should food webs used by a migratory organism differ isotopically and spatially, then stable isotope values in the consumer can provide unambiguous information on previous consumer locations.
2. The time period over which this spatial information is retained will depend on the tissue chosen. For metabolically active tissues, this represents a moving window of forensic information. For metabolically inactive tissues, spatial information will be locked in indefinitely but will only reflect position for the short period of integration reflecting the growth of that tissue.
3. Mechanisms related to dietary transfer of isotopic signals to consumer tissues including isotopic discrimination, exercise, and metabolic routing are known and accounted for.

In practice, it is rare that all three principles will be satisfied or known with sufficient confidence! However, depending on the organism, much of this uncertainty can be constrained and, as we shall see, useful inferences can be made with respect to previous provenance of individuals based on isotopic measurements of their tissues. The careful blend of knowledge of the life history of the organism of interest, knowledge of likely isotopic landscapes or "isoscapes" experienced by that organism, and the physiological parameters that can influence isotopic inferences makes up the *art* of using stable isotopes to track migratory organisms.

II. ISOTOPIC DISCRIMINATION

The "you are what you eat plus a few parts per mil" maxim is formalized in the equation:

$$\delta C_t = \delta d + \Delta_{dt}$$

where δC_t is the measured stable isotope value measured of a tissue in the consumer, δd is the stable isotope value of the diet, and Δ_{dt} is the diet-tissue isotope discrimination factor (Chapter 2). We now know that this discrimination factor is a great oversimplification and that it does not necessarily take into account metabolic routing of specific macronutrients such as proteins, lipids, and carbohydrates. Recent research has also determined that the diet-tissue discrimination factors are influenced by the quality of the diets and so are likely not static for most wild animal populations. Because our ability to place an organism in a particular isoscape is sensitive to our knowledge of the true discrimination factors associated with an organism or dietary regime of interest, researchers should bracket their estimates based on an honest assessment of how well they know such factors. At most, this will require dietary experiments with the organism of interest, or, at a minimum, a sensitivity analysis to determine the effect of varying discrimination values on the outcome of GIS models or other methods used to "place" an organism.

A. Nitrogen Isotopes

Stable nitrogen isotopes in the tissues of consumers really represent a means of tracing protein pathways derived from diet because this element is largely absent in lipids and carbohydrates. For essential amino acids, nitrogen will largely be incorporated with little isotopic discrimination into the protein pool of the consumer. Nonessential amino acids typically involve more opportunities for isotopic discrimination during protein synthesis and so the net discrimination we see for $\delta^{15}N$

measurements in consumers will reflect the degree to which the diet meets the amino acid requirement of the consumer (Robbins *et al.* 2005).

In general, poorer quality diets will likely result in greater overall diet-tissue discrimination for ^{15}N than high-quality diets. An important currency in experiments designed to establish tissue-specific $\delta^{15}N$ values in migratory animals is the C:N ratio of the diet and this ratio alone may provide a useful indicator of diet quality and the ultimate discrimination factor to apply in natural situations. Isotopic discrimination associated with $\delta^{15}N$ will also depend on the means of voiding nitrogenous waste. Here, a major difference is found between aquatic invertebrates that void nitrogen via ammonia compared to terrestrial vertebrates (Post 2002). There is also evidence that ungulates adapted to arid conditions are able to conserve water by recycling urea that ultimately influences whole body tissue $\delta^{15}N$ values (Ambrose and DeNiro 1986, Sealy *et al.* 1987). Hobson *et al.* (1993) also determined that birds that fast and undergo significant protein catabolism during incubation, like geese breeding at high latitudes, also experience an increase in body $\delta^{15}N$ values.

Knowledge of these sorts of physiological processes is necessary, when using tissue $\delta^{15}N$ values of migratory organisms to infer origins. The current consensus is that researchers should strive to use the most parsimonious value associated with their specific organism of interest. The review of isotopic discrimination in $\delta^{15}N$ across several taxa by Vanderklift and Ponsard (2003) identified mode of excretion and environment (marine, freshwater aquatic, terrestrial) as important factors (see also Post 2002).

B. Carbon Isotopes

Unlike nitrogen and sulfur, carbon is present in all three dietary macromolecules (protein, fat, carbohydrates) and so $\delta^{13}C$ measurements of consumer tissues will reflect these various sources. The more varied sources of carbon to consumer tissues undoubtedly contribute to more variable diet-tissue $\delta^{13}C$ discrimination factors compared with those found for most of the other light elements. However, in many cases, lipids in diets are transferred directly with little isotopic modification to lipids in the consumer. Carbohydrates are often burned directly for energy production, producing CO_2 as the only carbon byproduct, and hence $\delta^{13}C$ values in breath CO_2 can be used as a means of directly tracing origins of carbohydrates in diet. Unfortunately, we have little idea of the appropriate carbon isotopic discrimination factors that currently apply between dietary substrates and breath CO_2 in animals (Podlesak *et al.* 2005). Carbon isotope values of proteins can theoretically originate from all three dietary macromolecules but is more likely to be associated with dietary proteins, especially for carnivores. In general, we expect lower diet-tissue isotopic discrimination factors for $\delta^{13}C$ compared with $\delta^{15}N$.

C. Sulfur Isotopes

Sulfur in consumer tissues is derived from the sulfur-bearing amino acids (*e.g.*, cystein, methionine) and so $\delta^{34}S$ measurements are closely linked to dietary protein pathways. Unlike the other light isotopes, we expect little S isotopic discrimination between diets and consumer tissues again because of little opportunity for the essential amino acids to be isotopically modified in consumers. As a result, $\delta^{34}S$ measurements make for a useful direct tracer in food web and migration studies (Krouse *et al.* 1991, Hebert and Wassenaar 2005).

D. Hydrogen and Oxygen Isotopes

As emphasized below and in Chapter 4, hydrogen is a particularly useful element for tracking migratory wildlife. However, this element presents a number of challenges in terms of understanding how the δD measurements of consumer tissue relate to hydrogen sources that, in most terrestrial systems, is driven by the global water cycle, driving primary production. Like carbon, hydrogen occurs in all three dietary macromolecules and so recognition of metabolic routing is important. However, the most interesting challenge is the fact that a portion of the hydrogen in any tissue exchanges with body water, a component which is presumably more labile than dietary derived hydrogen. Drinking water as well as diet thus constitutes a source of hydrogen in animals. Using a controlled laboratory study, Hobson et al. (1999a) maintained quail (*Coturnix japonica*) on a single diet but exposed groups to drinking water of vastly different δD value. They found that hydrogen from drinking water accounted for about 20% of the total hydrogen in various tissues. Interestingly, this was the case for lipids with no exchangeable hydrogen bonds indicating that body water can exchange with hydrogen in precursor molecules involved in lipid synthesis. The overall diet-tissue discrimination factor for hydrogen is complicated to the extent that one ideally could apply the δD values of both diet and drinking water and knowledge of how these are partitioned for various tissues. Such details do not exist for most animal systems and so overall diet-tissue discrimination factors are currently estimated using more phenomenological approaches as discussed below.

The use of $\delta^{18}O$ measurements to track wildlife is in its infancy because of previous technological constraints of routinely measuring oxygen isotopes in animal tissues. That situation has now changed due to online pyrolytic techniques (Chapter 2). In many systems, there is a tight coupling between $\delta^{18}O$ and δD values because of the meteoric relationship described by Bowen and West (Chapter 4). Thus, in many cases, no additional information will be derived from performing $\delta^{18}O$ measurements in addition to δD measurements on the same tissue. As well, δD values typically span a much larger range than $\delta^{18}O$ measurements in terrestrial food webs and so can potentially provide greater resolution with respect to source discrimination and an overall better signal-to-noise ratio. Oxygen occurs in proteins but not in lipids or carbohydrates. However, sources of oxygen include drinking water and air and thus, like hydrogen, it is difficult to predict isotopic discrimination factors associated with each contribution and "working values" will need to be derived largely from future examination of wild and captive animals.

III. TISSUE AND ISOTOPIC TURNOVER: THE MOVING WINDOW

That stable isotope values in consumer tissues reflect an integration of feeding events over various time periods has been known and experimentally demonstrated for decades. Tieszen et al. (1983) were the first researchers to conduct "diet switch" experiments whereby captive animals were allowed to reach equilibrium under one dietary regime before being switched to an isotopically different diet (Figure 3.1). Tissues were sampled following the diet switch and the uptake of the new isotopic dietary signal monitored. This approach has now been used by several researchers examining various species and most have fit an exponential uptake curve to describe the pattern of isotopic change in tissues.

$$D(t) = a + b\exp(-ct)$$

where $D(t)$ is the stable isotope value of the tissue at time t, a is the asymptotic tissue value, b is the absolute change in tissue isotope value between initial and asymptotic conditions, and c is a rate constant defining tissue turnover. For cases where researchers wish to consider effects of growth (k) as well as metabolic turnover (m), the overall rate constant c can be expressed as ($k + m$). This approach has worked well to provide estimates of elemental turnover rates in various tissues of birds, fish, and mammals (e.g., Hesslein et al. 1993, Bosley et al. 2002, Dalerum and Angerbjörn 2005). One potential disadvantage of these previous studies is that they were necessarily based on sedentary, nonexercised individuals in laboratory settings. Can the elemental turnover rates obtained from such studies be

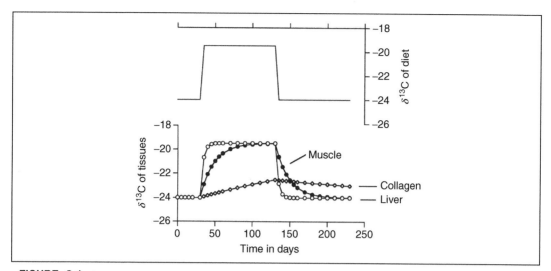

FIGURE 3.1 Conceptual depiction of the way in which different tissues will respond to an isotopic diet switch. We expect close coupling between the diet isotope trajectory and fast turnover tissues like liver or blood plasma. Much slower response is expected for slow turnover tissues like bone collagen. Based on the captive Japanese quail (*Coturnix japonica*) study of Hobson and Clark (1992).

directly applied to migrating individuals that, in the case of birds, can undergo many hours of sustained flight? Might we not expect more rapid elemental turnover in tissues of exercising versus sedentary organisms? This is still not clear but Hobson and Yohannes (2007) used Rosy Starlings (*Sturnus roseus*) trained to fly in a wind tunnel to provide a first approximation of this effect for the cellular fraction of blood. They performed a C3 to C4 diet switch on birds that flew several hours per day and contrasted the isotopic turnover rates with those of a control group, which was not exercised. Interestingly, they found no difference between the two groups. This suggests that erythrocyte production was unaffected by exercise at least to the level sustained in this experiment. More studies are needed and the use of wind tunnels is clearly the best way to explore turnover rates in migratory birds and insects. However, these results are encouraging and suggest that turnover rates established to date may be appropriate for most isotopic studies on migratory organisms. The other good news is that elemental turnover rates appear to follow expectations based on allometry (Figure 3.2). It is possible, then, to estimate turnover rates for various tissues based on the body mass of the organism of interest even though that species has not been tested experimentally (see also Carleton and Martínez del Rio 2005).

Recently, others have investigated a new way of analyzing and interpreting uptake curves based on isotopic dietary switch experiments (Cerling *et al.* 2007). Instead of fitting exponential equations to these curves to estimate turnover rates, they applied a technique established by radiochemists to estimate relative contributions of different radionuclide decay curves to a single (measured) decay function. This approach involves the determination of reaction progress variables that are derived from a process that involves linearizing the decay curves. Interestingly, such an approach has suggested that instead of representing a single elemental pool that changes according to the exponential function used previously, there is evidence that these curves can represent more than one source pool with each having a different elemental turnover rate. Perhaps the best current interpretation of this phenomenon is that essential amino acids are transferred relatively quickly from diet to tissues whereas nonessential amino acids are manufactured from dietary components and so represent a lag time prior to incorporation into consumer tissues. This new way of looking at elemental turnover in animals shows potential for promoting our understanding of how elements from diet and body stores are ultimately routed to consumer tissues and how these can differ temporally in terms of dietary integration.

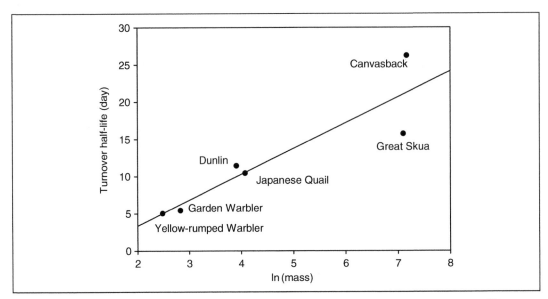

FIGURE 3.2 Relationship between turnover half-life (days) for whole blood as determined by a switch in $\delta^{13}C$ values of food of captive birds and the natural logarithm of body mass (grams). Based on studies by Hobson and Clark (1992), Haramis *et al.* (2001), Bearhop *et al.* (2002), Hobson and Bairlein (2003), Pearson *et al.* (2003), and Evans-Ogden *et al.* (2004). Figure shows the predominantly allometric relationship between elemental turnover rate and body size.

However, this important development in the way we consider turnover experiments in future by no means negates the results of earlier work using the more conventional approach and the *net* elemental turnover measured by fitting the exponential function still provides a phenomenological estimate of the time period a given isotopic measurement of an organism represents.

Once we have decided on a realistic estimate of the half-life of an element in the tissues of a migratory organism, we need to decide on a convention that best quantifies the time period represented by the isotopic measurement of that tissue. Most authors have considered that a tissue realistically represents about three half-lives or the time required for 87.5% of the original signal to be replaced by a new signal. Put another way, we should at least be able to detect the 12.5% of the original material remaining by our isotopic measurement. While this is a rule-of-thumb, our true ability to resolve between an original tissue signal and the asymptotic signal reached at a new location will depend upon the isotopic separation between these two signals and the nature of the distributions of these signals for populations of individuals in dietary equilibrium at each of these locations. The greater the isotopic difference between initial and final conditions (*i.e.*, the greater the value b), and the smaller the variance associated with each equilibrium condition, the more capable we will be of detecting isotopic information from a previous location (Figure 3.3).

As a hypothetical example, imagine a situation where the mean $\delta^{13}C$ value of a population at equilibrium with the diet at the new location (*i.e.*, a) can be statistically resolved at a minimum difference of 1 per mil from that of the more recently arrived population (such resolution will of course depend on the variance associated with each mean population isotope value, the nature of those distributions, and overall measurement error). If the population means were separated by only 4 per mil initially, then we would only detect a signal representing two half-lives of a given tissue. If, however, the populations were initially separated by 6 per mil, then we could detect a difference between these populations for three half-lives of that tissue. This example illustrates why researchers should strive to estimate the magnitude of these model parameters and the isotopic variance associated with them when trying to define the time periods over which their isotopic measurements correspond for migrating individuals or populations.

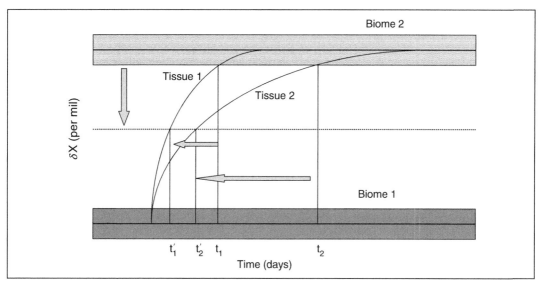

FIGURE 3.3 When an organism moves from one isotopic biome to another, our ability to detect the original biome signal will depend on the tissue we choose and the magnitude of the isotopic separation between the two biomes. Here, tissue 1 has a faster elemental turnover rate than tissue 2. The bands about each biome indicate the isotopic resolution or measurement error corresponding to organisms in that biome. The scenario of reducing the isotopic distance is demonstrated with the arrows. Here, we can see that the time over which we can detect the original biome signal is reduced (primed notation).

Phillips and Eldridge (2006) explored the utility of using more than one tissue for isotopic assay in order to detect the actual time an individual organism has spent in a new environment. Such information is less useful for estimating *where* an organism came from but can provide insights into the value of the migratory stopover environment. This approach is based on the contrast between tissues with different turnover rates, typically a fast turnover tissue like liver, plasma, or breath CO_2 and a slower turnover tissue like muscle or the cellular fraction of blood. The model assumes that the researcher knows the initial and asymptotic tissue isotope values, the measured isotope value of tissues at some time after arrival, and the necessary rate constants associated with the tissues used. The model does a good job of estimating the time since diet shift and the magnitude of the isotopic difference between initial and asymptotic conditions except in circumstances where the elapsed time was only a small fraction of half-life of the slower turnover tissue or when the diet shifts were small (*i.e.*, less than 10 times the measurement error).

As with all models, we are forced to make a number of assumptions to have any hope of achieving a solution. In the case of discerning dietary history or previous geographical provenance of a migratory organism, we typically make the assumption that the organism was in full equilibrium with its previous diet upon arrival. This will depend on the tissue used, and so there would be a much higher likelihood of equilibrium conditions reached for shorter turnover tissues. Use of intermediate- or long-turnover tissues may be less useful for a number of migrants that move quickly among different stopover locations. As we shall see later in examples of detecting dispersal in organisms, there are some circumstances where we are only interested in knowing that the arrival signature is *different* from the local food web signature and are less concerned that it can be associated with a particular location. There will also be error associated with estimates of rate constants $(k + m)$ and estimates of these errors are typically poorly known for most wild species. Finally, all current models assume a direct dietary source of nutrients to tissues and do not consider situations where organisms may be metabolizing stored nutrients for maintenance.

IV. APPLICATIONS OF STABLE ISOTOPES TO MIGRATORY MOVEMENT

The recent development of stable isotope methods to tracking migratory wildlife has already provided a rich literature to illustrate the breadth of applications using the light isotopes of C, N, H, O, and S. (reviewed in Hobson 1999a, Rubenstein and Hobson 2004, Hobson 2005a). In general, these applications can be split into (1) inferences of animal origins based on biome markers, typically using isotopes of C, N, and S and (2) those using continental-scale δD isoscapes. The use of δD measurements has brought with it immense opportunities but also challenges as we attempt to fill in the information gaps associated with this complicated element and so hydrogen will be discussed separately.

A. Migratory Movement Among Isoscapes

As it was recognized that stable isotope measurements can provide information on habitats or biomes, it was not long before researchers began to explore the use of stable isotopes to infer origins of migratory animals moving among such isoscapes. While the focus of this volume is terrestrial ecosystems, some of the earliest isotopic investigations revealed distinct differences between marine and terrestrial food webs with marine organisms typically having more positive $\delta^{13}C$, $\delta^{15}N$, $\delta^{34}S$, δD, and $\delta^{18}O$ values compared with their terrestrial counterparts (Hobson 1999a). As many migratory organisms use both terrestrial and marine biomes throughout their annual cycles, these marine versus terrestrial isotopic differences have become extremely useful (*e.g.*, Atkinson *et al.* 2005). Plant physiologists have also pioneered the use of stable isotope measurements to discern C3, C4, and CAM-based photosynthetic pathways using both $\delta^{13}C$ and δD measurements and have more recently investigated the effect of water-use efficiency mechanisms in C3 plants that generally leads to an enrichment of plant tissue ^{13}C.

One of the earliest applications of stable isotope methods to investigate animal spatial movement was made by Killingly (1980) who inferred the temperature of water during calcite formation of barnacles attached to the skin of California Gray Whales (*Eschrichtius robustus*) using $\delta^{18}O$ measurements and by Killingly and Lutcavage (1983) who examined $\delta^{18}O$ and $\delta^{13}C$ measurements in barnacles on loggerhead turtles (*Carettaq caretta*). That work has since been followed by a number of studies on the inorganic fraction of otoliths in freshwater and marine fish (*e.g.*, Meyer-Rochow *et al.* 1992, Kennedy *et al.* 1997). Other excellent examples of using marine isoscapes to infer spatial movements of marine mammals have been provided by isotopic analyses of the baleen plates of the western Arctic population of bowhead whales (*Balaena mysticetus*) migrating annually between the Beaufort and Bering seas (Schell *et al.* 1989) and southern right whales (*Eubalaena australis*) that annually cross the Southern ocean convergence, a zone of dramatic changes in food web $\delta^{13}C$ and $\delta^{15}N$ (Best and Schell 1996). Currently, there is great interest in establishing marine isoscapes that can be used to help track the movements of marine animals (Gómez-Diaz and González-Solis 2007, Hobson 2007).

Within terrestrial systems, some of the earliest applications of isotopic measurements to tracking origins of animals were conducted in Africa. That continent has attracted a number of isotopic studies over the years due to its importance archaeologically and as a center of current animal diversity and conservation concern. Terrestrial food webs in Africa also include varied C3- and C4-dominated isoscapes and, as discussed later, a very diverse and dynamic δD isoscape. Two simultaneous yet independent studies used stable isotope measurements of elephant (*Loxodonta africana*) ivory and bone collagen to infer origins of that material as a forensic tool to counter the illegal ivory trade (Van der Merwe *et al.* 1990, Vogel *et al.* 1990). Elephants feeding primarily on grasses sample a C4 food web and so have more positive $\delta^{13}C$ values compared to those feeding in woodlands on C3 browse. Elephants feeding in more arid areas may also have higher $\delta^{15}N$ values than those in more mesic habitats (Heaton 1987). Combined with assays of Pb and Sr isotopes, these studies showed strong segregation among several African elephant populations and underlined the forensic utility of stable isotopes to infer origins of several taxa. Unfortunately, some of the early enthusiasm was later tempered by the observation of strong year-to-year variations in food web $\delta^{15}N$ values within the Amboseli National Park presumably due to climatic variation (Koch *et al.* 1995). This illustrates the need to know the natural range of variation in stable isotope patterns spatially and temporally when evaluating the accuracy of the technique when inferring animal origins.

Fortunately, many terrestrial systems are less dynamic isotopically and strong and consistent isotopic differences are maintained over decadal or longer time frames. This is especially the case with the use of $\delta^{13}C$ measurements to track the use by animals of C3, C4, and CAM food webs. A very interesting application of $\delta^{13}C$ measurements to investigate mechanisms affecting the phenology of animal migration was that of Flemming *et al.* (1993) who showed that the nectarivorous bat *Leptonycteris curasoae* switched from C3 flowering plants during the winter to CAM flowering columnar cacti as it migrated north in the spring. The bat tissue isotopic data revealed how the species has adapted to the phenology of CAM "nectar corridors" during their northward migration. Hobson (1999b) showed that migrant insectivorous songbirds known to originate in boreal forest of North America had consistently lower feather $\delta^{15}N$ values than those species from more southern agricultural zones. Similarly, Hebert and Wassenaar (2001) were able to use $\delta^{15}N$ measurements to segregate waterfowl originating in agricultural zones that are typically enriched in ^{15}N compared to more natural temperate regions (see also Hobson *et al.* 2005).

A seminal study by Wolf and Martinez del Rio (2000) on the isotopic ecology of two species of doves using a desert region in the southwestern United States provides an excellent example of how tissue δD and $\delta^{13}C$ measurements can provide strong inferences on the diets, and hence feeding locations, of volant animals. The saguaro cactus (*Carnegiea gigantea*) represents a valuable food and water resource to animals in arid environments. This CAM resource is imbedded in a C3 plant biome and so fruit consumption can be traced by $\delta^{13}C$ measurements. However, water derived from CAM plants is also highly enriched in deuterium compared with surface waters and this provided evidence that while the two species consumed food from the saguaro, they differed in their dependence of this plant for drinking water. This topic will be revisited later in relation to factors affecting the use of deuterium basemaps for tracking migratory animals.

Other studies have exploited the strong C4 $\delta^{13}C$ signal of agricultural corn to infer the origins of migratory herbivorous birds. Alisauskas *et al.* (1998) were able to assign newly arriving lesser snow geese (*Chen caerulescens*) to a migratory staging area in the Missouri Valley, United States, because local birds subsisted largely on corn agriculture whereas immigrants arrived from distinct C3 habitats or non corn-growing regions to the south. In their study of feathers of red-winged blackbirds (*Agelaius phoeniceus*) across a latitudinal gradient in North America, Wassenaar and Hobson (2000) found that birds formed feathers from pure C3 to pure C4 food webs. Similar results were found for loggerhead shrikes (*Lanius ludovicianus*) on that continent (Hobson and Wassenaar 2001). Both of these examples indicate the way in which corn and other agricultural C4 crops imbedded in a C3 landscape can provide information on origins of migrant animals. However, $\delta^{13}C$ isoscapes in North America or other areas with intense agricultural production that have a mix of C3 and C4 plants will be necessarily complicated to model.

A study that clearly had profound influence on the field of identifying seasonal interactions or carry over effects from one season to another was that of Marra *et al.* (1998). Those authors examined the effect on tissue $\delta^{13}C$ values of American Redstarts (*Setophaga ruticilla*) occupying habitat of different quality on the Jamaica wintering grounds. Likely due to the effect of water-use efficiency in C3 plants, food webs in wet mangrove forest were more depleted in ^{13}C than those in xeric scrub habitat and such habitat markers were passed on to redstarts inhabiting those habitats. Long-term studies on this wintering population of redstarts suggested that those birds occupying the better (moister) habitat were the first to reach a body condition that allowed them to migrate back to the United States and

Canada to breed. Thus, the prediction was that early spring arriving male birds in New Hampshire would have muscle tissue relatively depleted in ^{13}C compared to later arriving birds. This was indeed the pattern that was found.

Why was this paper so well received (including the cover illustration of migrating redstarts on the cover of Science!)? Well, for years it has been known that early arriving birds on the breeding grounds tended to have greater reproductive success than those arriving later or initiating clutches later. It was long assumed that such effects on fitness were influenced entirely by factors on the breeding grounds. The Marra et al. (1998) paper clearly showed there was a direct link between conditions experienced by individuals on the wintering grounds and their ultimate fitness and renewed interest in the concept of migratory connectivity and seasonal interactions.

This work was followed up by an interesting modeling exercise by Norris et al. (2004a) who, using path analysis, were able to theoretically link tissue δ^{13}C values to number of offspring produced! Of course, this exciting work rests entirely on our ability to detect a wintering ground isotopic signal in newly arriving birds on the spring breeding grounds. Some have expressed concern that the use of whole blood or muscle tissue in small, migrating passerines is very risky because these birds may take as much as 2–3 weeks to reach their breeding destinations, and elemental turnover rates for these tissues are really quite rapid in these small creatures. Indeed, although statistically significant, the relationship between muscle δ^{13}C value and arrival date on the breeding grounds in the Marra et al. (1998) study was rather weak. Nonetheless, follow-up investigations have confirmed this winter ground signal using much larger sample sizes of arriving redstarts (Norris et al. 2005). Bearhop et al. (2003) also demonstrated the utility of using stable isotope measurements of slow-growing claw material in birds and intercepted Black-throated Blue Warblers (*Dendroica caerulescens*) during migration in the Bahamas Bearhop et al. (2004). Similar to the Marra et al. (1998) study, they found that birds in better body condition were from more mesic (moist) versus xeric (dry) habitats as shown by their claw δ^{13}C values.

Chamberlain et al. (2000) investigated δ^{13}C and δ^{15}N values in feathers of Willow Warblers (*Phylloscopus trochilus*) located along a breeding latitudinal gradient in Scandinavia. Two subspecies are segregated along this gradient that differ in morphology and assumed African wintering grounds based on a handful of band returns. Unlike most Nearctic migratory birds, Palearctic species typically replace flight feathers on their wintering ground in Africa and so it was possible to infer aspects of the wintering habitats using isotopic analyses of these feathers. It was found that the southern *trochilis* subspecies had feathers more indicative of the mesic habitats of sub-Saharan West Africa than the northern *acredula* subspecies with supposed wintering areas in Central, East, or South Africa. These results agreed generally with the limited band return data and with isotopic analysis of molting birds from known capture sites in Africa (Bensch et al. 2006). However, the follow-up study using a larger dataset by Bensch et al. (2006) found higher δ^{13}C values in general than those found by Chamberlain et al. (2000). This discrepancy is somewhat reminiscent of the forensic analysis of African ivory and the warning by Koch et al. (1995) that δ^{13}C and δ^{15}N isoscapes in some regions of Africa may be highly variable among years.

Because one of the principles in applying stable isotope methods to studies of animal migration is that we have direct or good knowledge of the isoscapes encountered by any organism of interest, it may come as a surprise to some readers that very useful information on the structure of migratory populations can be obtained even when this is not the case (Hobson 2005b). Barn swallows (*Hirundo rustica*) breeding in Denmark are known to winter in South Africa but more refined information on connectivity is not available. Møller and Hobson (2004) investigated δD, δ^{13}C, and δ^{15}N values in African-grown feathers of Danish swallows and discovered that the distribution of δ^{13}C and δ^{15}N values was bimodal with a rare group (6% of the population) having very different values compared to the main group. These researchers concluded that this represented different winter habitats or regions occupied in Africa by the two groups. They investigated other data obtained for these birds from the breeding ground and found that the "rarer" group had chicks of larger mass and tarsus than those of the common group and that their chicks also had a much reduced T-cell immune response, indicating their immune systems were less challenged. The current hypothesis is that this rare group in Denmark originated from breeding populations within Scandinavia. So, while no one would advocate a "shotgun" isotopic approach to the analysis of tissues from migrant animals, the structure of those distributions can also provide valuable information that can assist in the development of new testable hypotheses. The deliberate analysis of the isotopic structure of breeding populations of animals can also be used as a means of detecting immigrants into those populations. Hobson *et al.* (2004a) used this approach to investigate minimum estimates of dispersal into breeding populations of American Redstarts and Ovenbirds (*Seiurus aurocapillus*), using δD measurements of feathers.

Møller *et al.* (2006) continued their isotopic analysis of barn swallows in Europe by examining those populations occurring within and outside the zone of influence of the Chernobyl nuclear accident and for samples obtained before and after the incident. They determined that correlation between $\delta^{13}C$ and $\delta^{15}N$ values in winter-grown feathers differed significantly between regions for females but not for males. This was interpreted as indicating that birds from the non-Chernobyl region were composed of females from a smaller recruitment area. The reasoning was that individuals with greater dispersal distance molted their feathers across a larger wintering area, thus increasing the probability of higher isotopic variance. Weak positive $\delta^{13}C$ and $\delta^{15}N$ correlations within "isotopic populations" seem to be common and the decoupling of this relationship might be a good indicator of mixed populations. This area of research requires much more study. The intriguing nature of the Chernobyl study is that stable isotope methods might well provide insight into the occurrence of source and sink breeding populations of migratory animals, an aspect of populations that is often extremely difficult to evaluate by other means.

The movement of animals between marine, estuarine, and terrestrial or freshwater habitats holds great potential for inferring their past habitat use and potential migratory origins. Tietje and Teer (1988) were among the first to use stable isotope methods to investigate how wintering Northern Shoveler (*Anas clypeata*) ducks use coastal and inland freshwater wetlands and were able to demonstrate sedentary behavior among late wintering individuals. Other studies have primarily used $\delta^{13}C$ measurements to infer movement of piscivorous birds between marine and freshwater habitats (Mizutani *et al.* 1990, Bearhop *et al.* 1999), and Hobson (1987) even used this approach to infer use of garbage dumps by coastal wintering gulls near an urban center.

While not terrestrial, migratory movements of fish with an anadromous life stage present an isotopic opportunity and, as indicated, fish have the added advantage of carrying an isotopic record in their otoliths and scales (Nelson *et al.* 1989, Trueman and Moore 2007) and soft tissues (Hobson *et al.* 2007a). Kennedy *et al.* (1997) and Harrington *et al.* (1998) have nicely demonstrated how stable isotopes of several elements can be used on the organic and inorganic fractions of otoliths to identify natal streams of Atlantic Salmon (*Salmo salar*) intercepted as adults at sea. Essentially, the suite of $\delta^{13}C$, $\delta^{15}N$, and $\delta^{87}Sr$ measurements formed unique combinations of values reflecting the geological substrate and land-use practices surrounding drainage basins of key salmon-producing streams.

Even within entirely freshwater habitats in the terrestrial environment, there clearly is substantial isotopic variability that can be used to examine movements of fish and Hesslein *et al.* (1991) used $\delta^{13}C$, $\delta^{15}N$, and $\delta^{34}S$ measurements of muscle in broad whitefish (*Coregonus nasus*) and lake whitefish (*Coregonus clupiformis*) in two freshwater regions of the Mackenzie Delta in Northern Canada to infer their movements. Recently, there has been great interest in using fish tissue δD values because stream and river inputs to lake systems can have very different water δD values (Doucett *et al.* 2007). Thus, freshwater systems have great potential for a multi-isotope approach to trace migrations and movements of aquatic species. They also have the added advantage of being reasonably tightly constrained spatially and it should be possible to literally create multi-isotopic basemaps of the major aquatic space used by migrant fish.

B. Traveling to Breed: The Isotopic Tracing of Nutrients to Reproduction

A major impetus for studying migration is the concept that migration involves trade-offs between other life-history demands including reproduction, molt, and so on. The degree to which migratory animals carry with them nutrients that were acquired elsewhere for the production of young, the so-called *capital* (vs *income*) strategy is intrinsically linked to the costs and benefits of migration (Drent 2007). Although not an application of stable isotopes to infer origins of individuals or populations *per se*, the recent development of isotopic tools to trace origins of nutrients to reproduction is an important development that is of interest here (Hobson 2006).

As we have seen, a necessary principle to the application of isotopic models to infer migratory origins is the establishment of isotopic discrimination factors that link substrates to products. Here, the discrimination factors between endogenous (obtained locally) and exogenous (stored) sources of nutrients and eggs were needed to investigate such strategies in birds. Hobson (1995) raised Japanese quail (*C. japonica*) on a plant-based diet and peregrine falcons (*Falco peregrinus*) on a quail diet and then related dietary lipid and lipid-free $\delta^{13}C$ and $\delta^{15}N$ values in diet to corresponding isotope values in egg yolk lipid, lipid-free yolk, albumin, and shell carbonate. The derived isotopic discrimination factors between diet and egg components differed between the herbivore and carnivore diets reflecting, in part, differential macromolecular routing of lipids, carbohydrates, and proteins. An unexpected benefit of considering the carnivore income breeding model was the fact that it provided the first estimate of isotopic discrimination factors that were likely applicable to the capital breeding strategy. The reasoning here is simply that conversion of dietary protein (muscle) and lipids to eggs during an income process of egg formation should be kinetically and thus isotopically similar to the process of the production of eggs from endogenous muscle and lipid stores. Gauthier *et al.* (2003) were the first to apply this reasoning to generate the first fully quantitative estimate of the role of endogenous and exogenous nutrients to reproduction in Greater Snow Geese (*Chen caerulescens atlantica*).

The first study to use stable isotope measurements of egg components to infer origins of nutrients was Trust (1993) who examined $\delta^{13}C$ and $\delta^{34}S$ values in eggs of redhead ducks (*Aythya americana*) breeding in Manitoba, Canada. Hobson *et al.* (1997) measured isotopically eggs of three colonial waterbirds breeding on Lake Ontario, Canada, in order to determine if there was evidence for the transfer of marine nutrients acquired on the wintering grounds to eggs laid on the freshwater breeding grounds. That study lead to a more ambitious examination of five species of gulls, four species of terns, and one jaeger breeding on Great Slave Lake, Canada (Hobson *et al.* 2000a). As indicated by $\delta^{13}C$ and $\delta^{15}N$ measurements of tissues, birds arrived with largely marine-derived endogenous nutrients but only two species (*Larus argentatus*, *L. canus*) showed a strong indication of transferring marine-derived lipids to eggs and two species (*Sterna caspia*, *S. hirundo*) showed evidence of marine-derived protein contributions to eggs.

An interesting development in using stable isotopes to examine nutrient transfer to eggs was provided by the first use of hydrogen isotope (δD) measurements. Hobson *et al.* (2004b) examined δD and $\delta^{13}C$ values of endogenous reserves (muscle and fat) and egg components in redhead ducks breeding in Manitoba, Canada. δD measurements provide the advantage that not only is there a very large isotopic difference between terrestrial and marine food webs but that even among terrestrial food webs, deuterium can be a powerful indicator of latitude in North America (Chapter 4). The study showed striking patterns of isotopic change in endogenous tissues throughout the season for both isotopes but there was little evidence for any endogenous nutrient inputs to eggs for this species.

In a similar study using both $\delta^{13}C$ and $\delta^{15}N$ analyses of females and their eggs, Hobson *et al.* (2005) examined Barrow's Goldeneye (*Bucephala islandica*) breeding in central British Columbia, Canada. Although there was overall little evidence for a capital breeding strategy in goldeneyes, compared with later-laid eggs, the first-laid egg showed the greatest probability of including endogenously derived nutrients of both protein and lipid. Klaassen *et al.* (2001) provided a creative application of stable isotope analyses to nutrient allocations among Arctic waders by measuring $\delta^{13}C$ values in natal down of newly hatched individuals together with those values of feathers from adults growing feathers on the marine wintering grounds and juveniles later growing feathers on the breeding grounds. The authors reasoned that if eggs contained marine-derived endogenous nutrients, natal down should reflect that input and would thus be more enriched in ^{13}C compared to juvenile feathers that were presumably based entirely on local foods. Contrary to the inference of Klaassen *et al.* (2001), in their isotopic study of Red Knot (*Calidris canutus islandica*) and Ruddy Turnstone (*Arenaria interpres interpres*) breeding in the high Arctic, Morrison and Hobson (2004) found evidence for endogenous nutrient allocation to early laid eggs.

To date, the most quantitative attempt to ascertain the extent of endogenous and exogenous nutrient allocations to eggs has been conducted by Gauthier *et al.* (2003). That study used isotopic discrimination factors corresponding to capital versus income models in multisource isotope mixing models for protein and lipid contributions to eggs of Greater Snow Geese. Because the two groups of local foods available to geese, graminoids and forbs, were isotopically different, these authors used a three-input (*i.e.*, gramminoid, forb, endogenous tissue), two-isotope mixing model that accounted for differences in the [C] and [N] values of the foods.

C. The Deuterium Breakthrough

The success of the applications discussed so far depend to a large degree on how well we know the ecology of the species of interest (*e.g.*, does it use marine habitats for a portion of its annual cycle?, does it make use of corn or other isotopically distinguishable agricultural crops?) or on how well constrained or how well we know the isoscapes through which the animal moves or has access to. On a case-by-case basis, the success of a stable isotope approach could range between failures (*i.e.*, no *a priori* knowledge of the isotopic options available for inference) to providing unequivocal evidence of previous origins. The demonstration that deuterium and oxygen isotopes occur at continent-scale, and predictable patterns of abundance in precipitation and that those patterns could be passed on to animals growing tissues at those locations, provided a major breakthrough because this was the first example of a relatively robust isotopic basemap for use in animal migration studies.

As developed by Bowen and West (Chapter 4), deuterium values in precipitation in North America show a continent-wide pattern with a general gradient of relatively enriched values in the southeast to more depleted values in the northwest (Sheppard *et al.* 1969, Taylor 1974). Previous studies had also established strong correlations between growing season average δD values in precipitation and those in plant biomass (Yapp and Epstein 1982). However, it was the work of Cormie *et al.* (1994) on deer bone collagen that clearly demonstrated that such patterns were also passed on to organisms at higher trophic levels and that work inspired Chamberlain *et al.* (1997) and Hobson and Wassenaar (1997) to examine how well such precipitation isotope values were passed on to birds growing feathers at known locations at a continental scale across North America. The advantage to using birds was that their molt chronologies are relatively well known and most migrants breeding in the United States and Canada molt flight feathers prior to their southward migration. The ability to capture an individual on the wintering grounds and use its feather to predict an approximate latitude of breeding origin was indeed a major breakthrough and the strong correlation ($r^2 = 0.89$) measured by Hobson and Wassenaar (1997) between feather δD and the mean growing season average precipitation δD for forest songbirds across the central part of the North American continent was almost too good to be true. Fortunately, that large-scale spatial pattern has since been confirmed by several other researchers on various avian species (Table 3.1) even though some intriguing variation in this relationship has emerged (Chapter 5).

The first comprehensive application of the use of δD measurements in the study of migratory animals came fairly early with the investigation of the isotopic structure of Monarch butterflies (*Danaus plexippus*) wintering in Mexico. This example is worth considering in some detail because it

TABLE 3.1 Relationship between stable hydrogen isotope ratios of mean precipitation (δD_p) and the δD values of collagen, feather, hair, keratin, or chitin assumed to have been produced from those sources

Species	Equation	r^2	Model	Source
Birds				
6 species of North American songbird	$\delta D = -31 + 0.9\delta D_p$	0.83	H	Hobson and Wassenaar (1997)
6 species of North American songbird	$\delta D = -25 + 0.9\delta D_p$	0.88	B	Clark et al. (2006)
6 species of North American songbird	$\delta D = -19.4 + 1.07\delta D_p$	0.86	B	Bowen et al. (2005)
Black-throated Blue Warbler	$\delta D = -51 + 0.5\delta D_p$	0.86	CH	Chamberlain et al. (1997)
Red-winged blackbird	$\delta D = -27 + 1.1\delta D_p$	0.83	H	Wassenaar and Hobson (2000)
Bicknell's Thrush	$\delta D = -26 + 0.7\delta D_p$	0.48	H	Hobson et al. (2001)
Wilson's Warbler	$\delta D = -51.7 + 0.4\delta D_p$	0.36	B	Kelly (2000)
Wilson's Warbler	$\delta D = +14.47 + 1.41\delta D_p$	0.91	M	Paxton et al. (2007)
Wilson's Warbler	$\delta D = -21 + 0.7\delta D_p$	0.48	M	Meehan et al. (2004)
Mountain Plover	$\delta D = +17.4 + 1.26\delta D_p$	0.36	B	Wunder (2007)
23 species of European birds	$\delta D = -7.8 + 1.27\delta D_p$	0.65	B	Hobson et al. (2004d)
23 species of European birds	$\delta D = -22.3 + 0.77\delta D_p$	0.85	B	Bowen et al. (2005)
Cooper's hawk	$\delta D = -34 + 1.0\delta D_p$	0.83	H	Meehan et al. (2001)
Inland generalist raptors	$\delta D = -40 + 0.62\delta D_p$	0.59	H	Lott et al. (2003)
Inland bird-eating raptor	$\delta D = -44.2 + 0.54\delta D_p$	0.37	H	Lott et al. (2003)
Coastal generalist raptors	$\delta D = -38.8 + 0.55\delta D_p$	0.19	H	Lott et al. (2003)
Coastal bird-eating raptors	$\delta D = -104.7 - 0.59\delta D_p$	0.12	H	Lott et al. (2003)
Non-coastal bird-eating raptors	$\delta D = -41.1 + 0.58\delta D_p$	0.46	H	Lott et al. (2003)
9 species of raptors	$\delta D = -52.2 + 0.28\delta D_p$	0.09	H	Lott et al. (2003)
9 species of diurnal raptors	$\delta D = -37 + 0.6\delta D_p$	0.51	M	Meehan et al. (2004)
Raptors in South Carolina	$\delta D = -25 + 0.7\delta D_p$	0.18	M	Meehan et al. (2004)
Flammulated owl	$\delta D = -8 + 0.9\delta D_p$	0.66	M	Meehan et al. (2004)
12 species of raptors	$\delta D = -5.6 + 0.91\delta D_p$	0.62	M	Lott and Smith (2006)
Scaup	$\delta D = -27.8 + 0.95 D_p$	0.64	B	Clark et al. (2006)
Mallards and Northern Pintail	$\delta D = -57 + 0.835 D_p$	0.56	M	Hebert and Wassenaar (2005)
Other animals				
Deer collagen	$\delta D = 4 + 1.02\delta D_p$	0.94	C	Cormie et al. (1994)
Hoary bat	$\delta D = -25 + 0.8\delta D_p$	0.60	M	Cryan et al. (2004)
Monarch butterfly	$\delta D = -79 + 0.62\delta D_p$	0.69	H	Hobson et al. (1999b)
Beetle (chitin)	$\delta D = 33.2 + 1.60\delta D_p$	0.74	B	Gröcke et al. (2006a)

Model refers to basemap developed by Bowen et al. (2005): B; Meehan et al. (2004): M; the original or updated dataset used in Hobson and Wassenaar (1997): H; the dataset derived by Cormie et al. (1994): C, and the dataset derived by Chamberlain et al. (1997): CH.

still represents a good template for stable isotope tracking of migratory wildlife. The eastern population of the Monarch Butterfly in North America overwinter in about 13 known roost sites in the high-altitude Oyamel Fir (*Abies religiosa*) forests of central Michoacan and Mexico states. In spring, only gravid females migrate north, reaching Texas where they lay eggs on milkweed (*Asclepias* species) plants. The new generation emerging travels further north to repeat the process at higher latitudes. Finally, in one of the most spectacular migrations of any animal, in late summer only monarchs 4–6 generations removed form those ancestors that migrated northward from Mexico the previous spring then return to the *same* roost sites that they have never seen before.

Hobson *et al.* (1999b) first created an isotopic basemap corresponding to butterflies produced throughout their breeding range during the summer of 1996. An isotopic basement is composed of isotope measurements made on individuals from known locations that sufficiently spans the entire breeding range. This feat was accomplished through the aid of the nonprofit Monarchwatch organization (monarchwatch.org), who were able to solicit volunteers and educators from 86 locations across the monarch breeding range to successfully raised 4–12 butterflies on milkweed grown locally. Only milkweed watered by natural rainfall was used. From that sample, butterflies from 33 sites were selected for δD and $\delta^{13}C$ analyses of wing tissue performed to produce a year-specific basemap depicting isotopic patterns for C + H isotopes. In addition to the wild-reared group of monarchs, the relationship between δD and $\delta^{13}C$ of milkweed tissue and butterfly keratin and between wings and the water used to raise milkweed was investigated under controlled laboratory conditions using three batches of known δD water. Those captive studies showed extremely tight ($r^2 = 0.99$) relationships in each case demonstrating how insect wing keratin δD is derived exclusively from water available to plants with most of the isotopic discrimination occurring between water and plants (see also Ostrom *et al.* 1997). Wassenaar and Hobson (1998) then applied this basemap to portray origins of monarchs who were produced during 1996 and later collected from all known winter roost sites in Mexico that winter. That resulted in the insight that the winter roost sites were panmictic, made up of butterflies from all over the breeding range, and most importantly, revealed that half the population was produced largely in Kansas, Nebraska, Iowa, Missouri, Wisconsin, Illinois, Michigan, Indiana, and Ohio corresponding to the corn, soybean, and dairy producing region of the midwest. Thus, while conservation of this species was previously focused almost entirely on the precarious winter roosts in Mexico, these isotope studies pointed to the possibility that prime monarch breeding habitat was concentrated in areas of intense agricultural production in the United States where milkweed was controlled and where genetically modified corn was being used that produced BtK, a bacterium that targets Lepidoptera (Losey *et al.* 1999).

Here we revisit the original monarch dataset using modern GIS tools (see Chapter 4) compared to the original hand-kriged and interpolated results (Figure 3.4). Overall, we obtained a very similar finding to the original conclusions, but derived a more detailed picture of potential origins of monarchs that shows a somewhat wider potential band of origins (Figure 3.3). In this exercise, origins were assigned using georeferenced raster basemaps of expected δD and $\delta^{13}C$ surfaces interpolated from the original captive rearing work. Based on δD and $\delta^{13}C$ values from the wintering colonies in Mexico, 50% tolerance limits (TLs) at a 95% confidence level were calculated for each isotope separately (Walpole and Meyers 1993). Because assignment of origins is sensitive to the accuracy of the underlying surfaces, and/or individual variability at similar breeding locations, we calculated standard deviation of δD and $\delta^{13}C$ among individuals grown at the same location within the same rearing sites ($n = 140$ individuals at 33 sites). Average within-site standard deviation across all sites was 4.5‰ for δD and 0.3‰ for $\delta^{13}C$ values; these values were then used to extend the portrayed 50% TLs (*i.e.*, sensitivity limits) by subtracting these values from the lower TL and adding them to the upper TL. Origins were then portrayed for each isotope separately by reclassifying cells within the expected δD and $\delta^{13}C$ surfaces to values of 50% if they fell within the calculated TLs, and no data otherwise, using a spatial analyst reclassify operation within ArcGIS v 9.1 (ESRI, Redlands, California). In order to further constrain origins, the raster surfaces portraying 50% TLs (including sensitivity limits) for each isotope

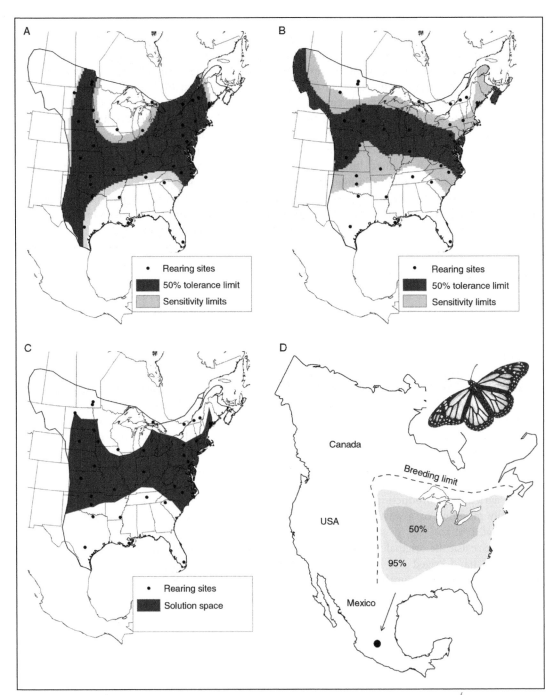

FIGURE 3.4 GIS-based natal origins of ~50% of Monarch butterflies overwintering population in Mexico using two stable isotope markers; (A) natal origins based on raster cells from an interpolated surface of $\delta^{13}C$ falling within the 50% tolerance limits (±0.29‰ (SD) to account for interindividual variability at sites); (B) natal origins estimated based on raster cells from an interpolated surface of δD falling within the 50% tolerance limits [±4.5‰ (SD)]; (C) the dual-isotope constrained estimate of natal origin for ~50% of the population using a GIS overlay of tolerance limits for both $\delta^{13}C$ and δD and incorporating both isotope's sensitivity limits. The results of panel C are therefore directly comparable to that of panel D; (D) the original hand-krigged and interpolated data of Wassenaar and Hobson (1998), using the original basemap calibration data of Hobson et al. 1999b.

were converted to vector format, and an ArcGIS intersect operation used to select only those portions of the two surfaces that overlapped. The way in which origins of populations are depicted using GIS tools is a work in progress but this exercise clearly shows the potential of this approach. Wunder and Norris (Chapter 5) further elaborate on Bayesian approaches that could be applied to deciphering origins of monarchs using our dataset.

The monarch isotope work continued several years later with an investigation into the source of butterflies occurring in Cuba in the autumn. There is a resident population of monarchs in Cuba, but researchers had noticed a seasonal increase in abundance. Stable isotope analyses were combined with cardenolide fingerprinting (related to the species of milkweed used and its geographical distribution) of monarchs and revealed that there was an influx of migrant monarchs to Cuba in November and that these individuals originated from southeastern Canada to the eastern United States (Dockx *et al.* 2004). So, it appears as though some monarchs make a "mistake" during their southward migration to Mexico and end up in Cuba. Recently, the origin of a single monarch that was found in Britain where they are considered an accidental vagrant was investigated. The mean wing keratin of this individual provided by Clive Farrell was -45 ± 2.5‰, clearly indicating that the butterfly was well outside the range of North American monarchs and was most likely from breeding populations off West Africa.

Notwithstanding the very clear conservation implications of the monarch isotope study, what made this such a powerful example of how to apply stable isotope methods to illuminate animal migration? Well, first, the monarch presents a very tight coupling between the animal of interest and its single genus of host plant. That clearly removed a source of variation in the isotope data for any regional population of monarchs. Second, the authors created a reference basemap using two stable isotopes that provided greater power to resolve origins of monarchs than could be obtained from either isotope used singly. Finally, the isotopic basemap applied exactly to the year of production of the cohort sampled on the wintering grounds and so removed variance associated with year effects or the use of long-term average precipitation datasets. In short, it was an organismal and year-specific basemap. Unfortunately, there will be very few opportunities for most of the other migratory animals of interest where we can control sources of variance and most studies will be forced to rely much more on inference based on long-term datasets such as the Global Network of Isotopes in Precipitation (GNIP). Having said that, a number of very important insights into origins of migratory animals have already been made using inferences based just on the pattern of deuterium in rainfall in North America and other continents, using the map lookup approach (Chapter 5).

Several recent applications using δD measurements have involved migratory birds in North America and have had a strong conservation motivation. Populations of loggerhead shrike have declined in North America and Hobson and Wassenaar (2001) and Perez and Hobson (2007) investigated the structure of wintering populations in the southern United States and northcentral Mexico. They were able to show that both Mexico and Florida were important wintering areas for northern breeding populations of this species. Because previous few band returns over the last 50 years pointed only to Texas, this work quickly identified new areas of potential concern. The Bicknell's Thrush (*Catharus bicknelli*) currently breeds in precariously small island habitats in the eastern United States and winters in the Caribbean. Based on a sample of birds wintering in the Dominican Republic, Hobson *et al.* (2004c) determined that a significant proportion of the population bred further north, likely in southern Quebec. Follow-up surveys indeed confirmed this. A similar study using both δD and $\delta^{13}C$ measurements of Black-throated Blue Warblers (*D. caerulescens*) on the wintering grounds revealed an important split in origins with those birds breeding in the south wintering in the eastern Antilles and those breeding in the north wintering in the west (Figure 3.3; Rubenstein *et al.* 2002). That southern populations seem to be declining compared with the northern populations suggests that deforestation patterns on the wintering grounds may be a factor. Other important advances, especially with migrant songbirds, have involved the delineation of catchment areas of constant-effort migration monitoring stations for a variety of species (Wassenaar and Hobson 2001, Mazerolle *et al.* 2005, Dunn *et al.* 2006).

By defining the approximate origins of birds caught during migration, analysis of population trends from these stations will be enhanced.

Another aspect of wildlife management concerns an understanding of where within the range of a species most of the young are produced. This is especially important for highly managed game species that are harvested during their fall migration and on the wintering grounds. Hobson *et al.* (2006) defined the origins of harvested Sandhill Cranes (*Grus canadensis*) through the central Flyway of North America using δD measurements of feathers and claws and identified the Hudson Bay lowlands and southern boreal forest as regions of highest productivity. Similarly, Hobson *et al.* (in review) examined origins of Lesser Scaup (*Aythya affinis*) taken by hunters during their autumn migration and compared isotopically inferred origins to *a priori* knowledge of the breeding densities of birds throughout most of their range. That exercise showed that more birds than expected were harvested from the central boreal and parkland region whereas fewer birds than expected were taken from the northern boreal. Such work has suggested that either there is a disproportional take of birds from the center of the species range and/or that the northern wetlands are producing fewer birds. The important point with both of these studies is that this type of information could not be obtained so readily using previous conventional mark-recapture techniques.

Other applications of the deuterium basemap for North America have been attempts to evaluate population structure and connectivity of migratory songbirds. By sampling birds across the wintering grounds, preliminary pictures are now emerging for Yellow Warbler (*Dendroica petechia*; Boulet *et al.* 2006), American Redstart (Norris *et al.* 2006), Henslow's Sparrow (Ibargüen 2004), loggerhead shrike (Chabot *et al.*, unpublished data), and Mountain Plover (Wunder 2007). These studies are primarily limited by the amount and geographical extent of winter captures. Kelly *et al.* (2002) also nicely demonstrated a "leapfrog" migration system in western populations of Wilson's Warbler (*Wilsonia pusilla*; Figure 3.5).

An interesting aspect of the behavior of δD is that deuterium in precipitation tends to rain out more at lower elevations than at higher elevations. This is a well-known phenomenon that results in an altitudinal "depletion" in δD from -1 to $-4‰$ per 100 m rise in elevation, depending on the gradient and temperature change (Clark and Fritz 1997). Similarly, the demands of plant adaptations to harsher

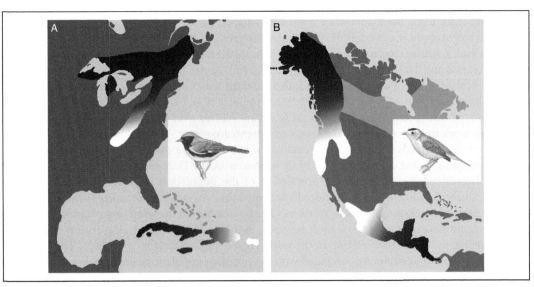

FIGURE 3.5 Migratory connectivity determined using stable isotope analyses of feathers based on the results of (A) Rubenstein *et al.* (2002) for the Black-throated Blue Warbler (*Dendroica caerulescens*) and (B) the leapfrog migration pattern of the Wilson's Warbler (*Wilsonia pusilla*) discovered by Kelly *et al.* (2002). **(See Color Plate.)**

growing conditions at higher altitudes tend to result in plants with higher $\delta^{13}C$ values at higher elevations. As there are several species that perform altitudinal migrations, especially in the tropics, it was of interest to see if tissues grown along an altitudinal gradient reflected such patterns. Hobson *et al.* (2003) examined δD and $\delta^{13}C$ values of feathers of hummingbirds inhabiting the Ecuadorean Andes and found good agreement between actual and predicted feather δD based entirely on a global model (see Chapter 4). As expected, feather $\delta^{13}C$ values increased with altitude. Thus, for any given species, it should be possible to estimate approximate elevations at which feathers and other tissues were grown. By examining different tissues with different windows of isotopic integration, the possibility exists to infer previous altitudinal movements.

To date, primarily applications to the North American continent have been presented. That is understandable since the first studies using animal δD measurements were conducted there and the pattern of δD in precipitation follows a strong gradient with latitude. More recently, applications have been developed in Europe and Africa. The first depiction of the European growing-season basemap was presented by Hobson (2003) and since then more sophisticated treatments have been presented by Bowen *et al.* (2005). A first test of the relationship between feathers and predicted precipitation δD in Europe was presented in Hobson *et al.* (2004d) and that study revealed little difference between the relationships derived using growing season or annual precipitation δD. In addition, a poorer relationship was found with these regressions compared to similar work conducted in North America. Nevertheless, some valuable work has already been published based on the European picture.

Bearhop *et al.* (2005) used feather and claw δD measurements to distinguish between those Blackcaps (*Sylvia atricapilla*) breeding in Germany that had wintered in the United Kingdom and those that had wintered in Spain. The overwintering of this species in the United Kingdom is a relatively recent phenomenon encouraged by changes in climate and the provisioning of overwintering birds by feeders. The consequences of this "new" wintering population was that compared with the more traditional Spanish wintering birds, those wintering in Britain arrived in Germany earlier and tended to mate with others from the same wintering grounds. This "assortative mating" provides the conditions for speciation and was elegantly quantified using only δD measurements. Other investigations in Europe have inferred the origins of irruptive species to western Europe like Bullfinches (Newton *et al.* 2006) and, more recently, Crossbills (*Loxia curvirostra*; Marquiss *et al.* in press). The crossbill study was able to contrast origins of individuals from several eruptions to Scotland over a 100-year period using feathers from museum specimens. More recently, Lormee, Hobson, and Wassneaar (unpublished data) are investigating the origins of hunter-killed wood pigeons (*Columba palumbus*) in France using feather δD measurements. That study is designed to establish important baseline information to help manage the hunt of this species in Europe. Researchers at the University of Lund are also using δD measurements to investigate migration in two species of butterfly (*Vanessa atalanta*) occurring in Europe (O. Brattström and S. Åkesson, unpublished data).

The deuterium basemap for Africa shows rather dramatic changes seasonally. However, an interesting and potentially very useful feature is the more depleted values in the southern part of the continent and the extremely enriched region in the northeast, centered on Sudan and Ethiopia. In their investigation into potential wintering sites of the endangered aquatic warbler, Pain *et al.* (2004) did not find δD measurements to be particularly useful and instead advocated the use of $\delta^{15}N$ and $\delta^{13}C$ measurements to define wintering areas in Africa. On the other hand, Yohannes *et al.* (2005, 2007) has investigated δD together with $\delta^{15}N$ and $\delta^{13}C$ measurements in feathers of several migrant passerines moving through East Africa where some of them stop to molt en route to more southern wintering areas. In these studies, δD measurements were useful in segregating among locally molting birds and those that delayed their molt until reaching their final (southern) destinations. One of the challenges to applying the δD approach in Africa will be to determine which rainfall matters in terms of regional and local food webs and it may well be that depictions of mean annual precipitation δD isoscapes are less useful than those based on more seasonal periods.

While the vast majority of studies tracking migrant animals using the deuterium approach have been on birds and insects, Cryan *et al.* (2004) applied this technique to migratory bats and produced a regression between altitude-corrected mean growing season precipitation δD and δD of hair from known site of capture where bats presumably grew their hair. That study was based on museum specimens taken over a 97-year period but there was no evidence of effect of time on the regression. The reasonably good regression ($r^2 = 0.6$) suggests that while mammal hair is composed of α-keratin and bird feathers are made up of β-keratin, such structural differences have little effect on discrimination during hydrogen uptake ultimately from precipitation. Cryan noted that other sources of variation in the dataset may have been because of effects of heat stress (McKechnie *et al.* 2004), relative humidity (Cormie *et al.* 1994), sources of water to food webs other than precipitation, and possibly northward postmolt movements of bats away from their molt sites where they were subsequently captured.

D. The Deuterium Challenge

Despite the tremendous advance provided by the deuterium basemaps now available for several continents, there are a number of issues related to this isotope that need to be resolved before the field can advance further. Some of these issues undoubtedly relate to the fact that CFIRMS techniques have only recently become available for δD and to date, no official organic standards are available to help coordinate measurements and quality control among laboratories. While these issues can largely be resolved using comparative equilibration techniques discussed in Chapter 2, there are still laboratories that are not following this protocol. It must be stressed that reproducibility can be achieved routinely within $\pm 2‰$ for keratin material like feathers and hair. Also, by using reliable analytical techniques, researchers will not need to rely on running all of their samples in a narrow window of time to avoid effects of ambient water vapor in the laboratory and their measurements will be comparable among all time periods and all laboratories participating in this approach regardless of location. Such an approach would reduce the necessity of applying dubious "correction factors" among locations where samples were measured (*e.g.*, Norris *et al.* 2006).

Beyond issues related to measurement error, there is the issue of the appropriate precipitation to feather (or other keratin) δD discrimination factor that should applied to our data in order to create expected keratin δD surfaces or basemaps (for those cases where a species-specific basemap like that of Lott and Smith (2006) does not exist). For songbirds, a value of $-25‰$ has been suggested (Hobson 2005a). That value was based on measurements of ovenbird (*S. aurocapillus*) feathers grown by adult birds that returned to a study site over several years and insights provided by measurements of captive red-winged blackbirds. That value also seemed appropriate from a regression of feather δD against growing-season average precipitation δD for birds sampled across a large gradient in latitude in North America (Clark *et al.* 2006). Recently, Langin *et al.* (2007) provided additional support for this discrimination factor based on δD measurements of adult American Redstart feathers of known origin. However, such a discrimination factor is almost certainly not applicable to all birds in all situations (Table 3.1) and Wunder and Norris (Chapter 5) discuss in more detail the sources of variance in these regression relationships, including slope and y-intercept. In their analysis of songbird feathers from Europe, Hobson *et al.* (2004c) derived a different relationship between feathers and growing-season precipitation δD, albeit based on birds where origin of feather growth was only assumed and not proven. Also, in their extensive survey of raptors across North America, Lott and Smith (2006) determined that their estimate of precipitation to feather δD discrimination value was generally greater in magnitude than that derived for songbirds and showed evidence of geographical variation. Areas of North America showing the greatest departure (*i.e.*, the highest residuals for the regression between measured feather δD and precipitation δD) of raptor feather δD values from expected were along the Pacific coast of the continent and in the arid southwest region of the United States. Lott and Smith speculated that the coastal effect may have been due to departures of the predicted precipitation δD

because of dynamic weather effects, not modeled well in this region. They also suggested that the deuterium enrichment in feathers found for the arid southwest region may have been due to the incorporation of CAM plants into the local food webs used by raptors. Another explanation is that this region of high residuals corresponds very well to that influenced by the North American monsoon, an area of northwestern Mexico and the southwest United States that often experiences significant late-summer rains, and it is quite possible that these differentially drive local food webs used by molting raptors and other species. This underlines the importance of considering regional "problematic" areas when using isoscape basemaps (Chapter 4) and also points to regions of North America that undoubtedly require more research and ground truthing.

In some ways it is not at all surprising that estimated tissue to precipitation discrimination factors for δD will vary and that the relationships measured so far rarely have a slope of unity. First, hydrogen is an element that will exchange with weak O–H or N–H bonds and this can take place with drinking water and overall body water (Hobson *et al.* 1999a). Second, we know that body δD values can increase as a result of heat stress (McKechnie *et al.* 2004) and presumably as a function of work or high metabolism that results in increased body evapotranspiration. Powell and Hobson (2006) found that Wood Thrush (*Hylocichla mustelina*) growing feathers in Georgia had higher feather δD values than expected from the feather δD basemap and speculated that heat stress during molt may have been a factor. Third, while we assume that trophic δD discrimination effects are minor and that most of the precipitation to tissue discrimination occurs between precipitation and plants, this has not yet been investigated experimentally (Birchall *et al.* 2005). Notably, it is indeed possible that feathers of birds grown in the nest may differ from those of the adults feeding them due to differences in metabolism, drinking water, thermal regime, and diet. Langin *et al.* (2007) found evidence that feathers of young redstarts indeed differed from those of adults.

An impressive number of studies have examined the relationship between animal tissue δD and the predicted precipitation δD values corresponding to real or assumed origins (Table 3.1). The amount of variance explained by these regressions range from 36% to 91% for cases not influenced by possible marine sources of hydrogen (Lott *et al.* 2003). However, it is also clear that there is considerable variation in the nature of these regressions even within groups like songbirds and for similar tissues (*i.e.*, feathers). Unfortunately, these relationships encompass all sources of error and it is currently not clear what is driving such differences. Several studies differ in the lab protocols used and the way in which they have dealt with equilibration through the use of various keratin standards (Chapter 2). It is also possible that the models used to estimate the expected mean growing-season average precipitation δD influence the results even though they are typically based on the same GNIP dataset. The range in regression relationships seen for three independent studies of the Wilson's Warbler in the western region of its range in North America is particularly interesting. Again, this topic receives more in-depth treatment in Chapter 5.

Another concern among would-be users of the deuterium basemaps for the various continents is the variability inherent in using a 40+-year average IAEA GNIP dataset for any given year where organisms are sampled (Chapter 4). As demonstrated with the monarch study, the only sure way to avoid this sort of variance is to create a basemap for the year of interest but this is clearly beyond the scope of most researchers for most organisms. Second, if a study site is close to one of the IAEA GNIP sampling stations, then it is possible for the researcher to potentially derive a year-specific tissue value for the site of interest but this would only provide local information most appropriate for discerning local from immigrant individuals (Hobson 2005a). More realistically, if researchers can obtain animal tissues from known individuals grown in the year of interest that could act as a reasonable proxy for local integrated isotope values.

The following chapters will address more specifically the nature of the available online products now available for researchers to estimate mean annual or mean growing-season precipitation δD values for their sites. Those products use an average monthly precipitation surface to weight contributions to long-term annual or growing-season precipitation δD. While this is entirely appropriate and these products have been of immense assistance, researchers need to be aware that changes in long-term weather patterns can result in departures of real average precipitation δD from those predicted from the model, a concern in these times of potentially rapid climate variation. Areas such as Saskatoon, Canada with its own GNIP sampling site, that experienced drought conditions in the last decade have shown differences between measured and assumed growing-season precipitation δD as much as 20–30‰.

Another issue involves our general poor understanding of which rainfall matters. The good relationship obtained between feather δD and mean annual growing season δD in North America (Hobson and Wassenaar 1997) was for *forest* birds distributed through the central region of the continent. Closed-canopy forest with shallow root systems may well integrate food web δD available to birds and other animals over such long time periods. However, is this the case for more pulsed ecosystems like grasslands, or deserts? In arid areas, rainfall driving local food webs can occur in just a month or two and different seasonal rains can drive different components of the food web as mentioned above with the North American monsoonal rains in southeast United States and northwest Mexico (Ehleringer *et al.* 1991). In other riparian systems, snowmelt may have the greatest influence on local food web δD. In other systems where animals may be influenced by aquatic emergent insects, tissues grown later in the season may be more enriched than those grown earlier if evapotranspiration enriches aquatic food web δD over the season. Much to our surprise, waterfowl feather δD values follow closely the expected growing-season average value for at least the temperate region of North America (Clark *et al.* 2006), despite the potential for this enrichment effect. All of these sources of variation require careful consideration.

There have been some interesting papers published over the last decade that have called into question the general applicability of using feather δD measurements to infer origins of migratory birds and it is worth considering a few of these carefully here. Wunder *et al.* (2005) provided a very useful contribution in their study of Mountain Plover (*Charadrius montanus*) chicks across their breeding range in western North America. They found that δD, $\delta^{13}C$, and $\delta^{15}N$ measurements were poor predictors of latitude using probability-based models (see also Chapter 5). Of particular note was the high range of chick δD values for a given site. One can only speculate on the factors responsible for this. As noted above, growing chicks may be under a very different thermoregulatory and metabolic regime compared with adults and this may have consequences for their body δD values. In this regard, we may also find differences between chicks of precocial versus altricial species but that has not been investigated. Diet and the source of water to local food webs may have been highly variable in these habitats and range from snowmelt to precipitation. Of note also in this study (although not related necessarily to within-site isotopic variance) is the fact that latitude per se may not be the best metric to use in a part of the continent where isotopic δD contours are convoluted due to altitudinal effects. However, regressions involving feather δD versus predicted mean growing season δD for both the Mountain Plover dataset and the Wilson Warbler (*W. pusilla*) dataset of Kelly *et al.* (2002) actually show lower r^2 values compared with those using latitude (Wunder 2007, J. Kelly, pers. comm.).

Another paper using δD together with $\delta^{13}C$ and $\delta^{15}N$, analyses involving shorebirds was that of Rocque *et al.* (2006) who examined feathers of American and Pacific Golden Plovers (*Pluvialis dominica* and *P. fulva*) and the Northern Wheatear (*Oenanthe oenanthe*). All species bred in Alaska where they were sampled and wintered in South America, the Pacific islands and Asia, and Africa, respectively. The authors expected winter-grown feathers to be much more enriched in deuterium compared with those grown in Alaska based on generalized patterns of δD in precipitation throughout the globe (Bowen *et al.* 2005). The authors also felt there was enough *a priori* information to expect that the isotope approach would successfully discriminate these birds to known groups based on their continent of molt. Wheatears behaved as expected with Alaska-grown feathers agreeing closely with that expected from the deuterium feather basemap. However, winter-grown wheatear feathers were not enriched in deuterium as expected and in fact were not different form the Alaska feathers. Variation among feathers of *P. dominica* was large, and those of *P. fulva*, much more enriched than expected. Wintering ground feathers of the plovers were more enriched than the breeding grounds but were also highly variable. The authors claim to be confident that the feathers were grown where they assumed and so what to make of these data? Like the Wunder *et al.* (2005) results, shorebirds seemed to show unexpectedly high isotopic variation within a single (100 km) location in Alaska. As nonstop migrants, often traveling thousands of kilometers in a single flight, these birds are uniquely adapted physiologically to turning themselves from feeding machines to flying machines with truly amazing adaptations of shifting tissue resources (including digestive tracts) (Piersma *et al.* 1999, Battley *et al.* 2000). Perhaps such adaptations influence variation in feather δD values. Shorebirds may also sample a broad range of terrestrially linked food items that show microgeographic isotopic variance. Clearly, we need to know more about this group using captive experiments. As for the Wheatears, is it possible they grew those winter feathers on the breeding grounds? If not, then is it also possible that they grew feathers in Africa based on high-lipid (and thus with depleted δD values) diets in anticipation of a nonstop trans-Atlantic flight? Finally, with the complex array of isoscapes available to birds on wintering grounds at *continental* scales, especially those for shorebirds that use everything from marine to high altitude terrestrial biomes and agricultural areas, is this study really a test of the stable isotope approach to migration studies? Alexander *et al.* (1996) found large isotopic ($\delta^{13}C$ and $\delta^{15}N$) variability within a single stopover site used by migrating shorebirds in Saskatchewan, Canada.

In addition to shorebirds, another "problematic" group of birds from a deuterium isotope perspective are the raptors. Despite some early encouraging results from Meehan *et al.* (2001) in depicting origins of migrating Cooper's hawks (*Accipiter cooperii*) through southern Florida, more recent studies have pointed to some unexpected results. As mentioned, Lott and Smith (2006) have generated a raptor feather basemap for North America based on collections of feathers at nests and on museum samples. That product is certainly the most extensive available for any animal group and points to likely differences between passerines and raptors in the precipitation to feather δD discrimination factor. Meehan *et al.* (2003) in their contrast of nestling and adult Cooper's hawk feathers grown at known sites found a good agreement with expected for nestlings but a poorer fit for provisioning adults. While this may have been related to dietary differences, others have suggested that adults molt during the brood rearing phase and so may become more enriched in deuterium than expected due to this extra work phase (Smith and Dufty 2005). Most birds separate molt and regrowth of feathers from other energetically taxing periods like breeding and migration.

The potential linkage between feather δD values and factors related to workload and physiology is a fascinating one. In their much celebrated study on American Redstarts (resulting in an unprecedented *second* cover illustration of redstarts for Science!), Norris *et al.* (2004b) concluded that males investing in reproduction late in the breeding cycle faced a trade-off in subsequent timing of molt and migration and were forced to molt tail feathers during migration. Thus, individuals whose breeding schedule was known the year before and who tended to breed late had tail feathers more enriched in deuterium suggesting molt south of their breeding grounds. In addition, tail feathers corresponding to these more southern locations had fewer carotenoids present, an independent measure of stress during feather growth (Norris *et al.* 2007). The reason this paper was so important is the fact that it provided one of the first independent measures of a trade-off between reproduction and other life-history traits faced by a migratory organism. However, there are other possible interpretations that now require further investigation. First, birds were forced to breed late due to experimental manipulations and so these data were not based on entirely natural circumstances. Second and most important, the physiological stress encountered by late breeding, after several previous nesting attempts, may have simply elevated the body δD values such that feathers molted on the breeding ground were correspondingly enriched in deuterium compared with birds less taxed (see Norris *et al.* 2007). Perhaps, redstarts were "changed into raptors" in this respect.

In another study on redstarts at the same study site, Langin *et al.* (2007) examined the range of variation in redstart feather δD values for birds known to have attempted breeding at the site the year before and over a 4-year period. Those authors cautioned that the *range* in values they measured should be a warning to researchers interested in using the δD approach to depict origins of birds in

North America. However, their study stands out as a shining example of an almost exact fit between the mean feather isotope value of their entire population over 4 years and the *expected* value based on the application of a −25‰ discrimination factor between mean growing season precipitation δD for their site and feather δD. Moreover, of the 42 individuals sampled, half were within 2‰ and 80% were within 6‰ of the mean. This now begs the question of what was driving the few outliers and how should we deal with these sorts of questions. According to the paper by Norris *et al.* (2004b), some of those outliers could have been birds that molted further south of this breeding site. On the other hand, Langin *et al.* (2007) measured a large range in insect prey δD values at this riparian site, so it is equally possible that food δD may change seasonally here and birds growing feathers late sample a different isotopic food web than those growing feathers earlier. All of which is to say that we often require a "reality check" when considering the constraints we should place on our isotopic interpretations, especially when we do not know well what may really be going on in nature. Langin *et al.* (2007) placed a great deal of emphasis on the *range* of feather δD values they measured rather than emphasizing the tight nature of the distribution. This pertains to the risk of incorrectly assigning locations to *individuals* (see also Wunder *et al.* 2005, Rocque *et al.* 2006, Wunder 2007, Chapter 5). Another approach, where appropriate, is to consider defining origins of *populations*. It is precisely that approach that has been taken by authors who have used GIS tools to depict 50% or 75% TLs of origins of avian populations or other means of delineating origins of portions of the population (Wassenaar and Hobson 1998, DeLong *et al.* 2005, Hobson *et al.* 2006, 2007b, Lott and Smith 2006, Figure 3.2).

The paper by Norris *et al.* (2004b) underlines the need to know what life history factors can contribute to interpretation of tissue stable isotope values. For birds, molt patterns are reasonably well known for most species. However, stable isotope measurements themselves have provided important qualifiers. The molt of flight feathers of northern populations of the loggerhead shrike are essentially bimodal with inner primaries, secondaries, and tail feathers usually being molted on the breeding grounds but other feathers being grown on the wintering grounds following a suspension in molt, a pattern discovered accidentally using stable isotope measurements (Perez and Hobson 2006). Other birds undergo prealternate molt of some body feathers on the wintering grounds prior to migration allowing us to investigate aspects of winter origins or habitat use (Mehl *et al.* 2005, Mazerolle *et al.* 2005). Unfortunately, information on the reliability or extent of prealternate molt or on the extent of delayed molt in migrating birds is often not available (Hobson *et al.* 2000b). Another alternative is to use claws that are continuously growing. Birds captured soon after their arrival on the breeding grounds should have claws that have retained information from the wintering grounds (Bearhop *et al.* 2005, Mazerolle and Hobson 2005). While we need more controlled studies to establish growth rates of claws for a variety of species, contrasting stable isotope values of claws against a metabolically active tissue like blood can in fact provide insight into periods where these tissues "agree" isotopically.

V. SUMMARY

This chapter started with the admission that situations where all three principles of isotopic tracking of migratory animals are met will be rare. The degree to which researchers are successful in applying isotopic methods will depend very much on the organism of interest, its geographical range, and ecophysiology. Such applications also fundamentally depend on how well we know the nature and behavior of the appropriate isoscapes. The most elegant applications will usually be situations where alternative isoscapes are very different and species experience simple isotopically dichotomous situations during their travels. Here, the long-standing success in using stable carbon isotope analyses to delineate C3 versus C4 or CAM food webs or the use of stable hydrogen isotope analyses to further separate C4 and CAM pathways provide distinct advantages. Terrestrial organisms that also spend part of their lives in marine or estuarine situations definitely lend themselves to isotopic tracking using several elements. Altitudinal migrants constrained by latitude and longitude also represent a useful application of δD and δ^{18}O measurements providing the movement represents several hundred meters. We will have more trouble in cases where underlying isotopic gradients are less distinct or where alternative origins overlap isotopically.

The application of deuterium measurements in animal tissues to place them on continental basemaps undoubtedly has provided the single greatest impetus in this field of isotopic tracking. Again, the success of this approach will depend very much on which part of the basemap we are dealing with. Distinguishing between arctic and prairie origins of migratory birds in North America or between those from Scandinavia or Spain in Europe will be relatively straightforward. We are faced with more of a challenge in distinguishing between birds or other animals originating across latitudinal bands on both continents or from regions that are more spatially restricted (*e.g.,* Szymanski *et al.* 2006). So, while this chapter has shown that the isotope approach has provided an extremely exciting and powerful boon to researchers and conservationists, it is not a "silver bullet" to be applied without full recognition of the limitations. How then might the field proceed from here? The next two chapters will deal explicitly with the very dynamic fields of isoscape mapping and statistical inference as they relate to placing organisms to origin and it is these areas that very much deal with key quantitative constraints we face. We now realize that the application of a single precipitation to tissue δD discrimination factor across diverse species or particular geographic regions will inherently involve error and sensitivity analyses should be part of our approach in future studies (*e.g.,* Figure 3.4; Hobson *et al.* 2007b). In addition, we should realize the risks in placing all of our eggs in one isotopic basket. For example, several researchers have used three or four isotopes (δD, δ^{34}S, δ^{13}C, δ^{15}N, δ^{87}Sr) in a multivariate approach to examine origins of migratory populations (Caccamise *et al.* 2000, Hebert and Wassenaar 2005, Yohannes *et al.* 2007) but more isotopes is not necessarily a guarantee of greater spatial resolution. The particularly elegant study of migratory raptors by Lott *et al.* (2003) is also instructive. Those authors nicely separated coastal migratory raptors having access to marine-based resources from inland species using δ^{34}S measurements of feathers. This approach will theoretically allow the subsequent identification of those migrant birds where δD measurements could be used to approximate origins using the appropriate terrestrial δD isoscape (Table 3.1, and Figure 3.6).

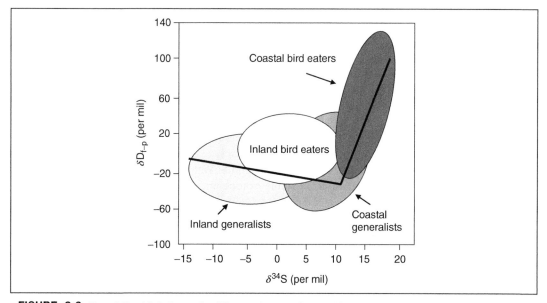

FIGURE 3.6 The relationship between the difference between feather δD and predicted precipitation δD (δD_{f-p}) and feather δ^{34}S for nine species of raptors breeding in North America (from Lott *et al.* 2003). This figure illustrates the way in which birds having access to marine protein can be distinguished by their enriched δD and δ^{34}S values.

By this point, the reader should be encouraged by the breadth of past isotopic applications to tracking migratory animals and realize the tremendous scope for future developments. The need for caution and consideration of the numerous assumptions involved will be sobering (Chapter 5). Nonetheless, more and more we are coming to terms with the nature of isotopic variance in the natural world and Kelly *et al.* (Chapter 6) provide direction on the path ahead.

VI. ACKNOWLEDGMENTS

Steve Van Wilgenburg and Kathryn Pieman assisted with illustrations. Dustin Rubenstein provided earlier drafts of Figure 3.4. Len Wassenaar provided excellent comments on a previous draft of the chapter. Mike Wunder, Stuart Bearhop, Ryan Norris, and Jeff Kelly provided useful reviews and/or unpublished data.

VII. REFERENCES

Alexander, S. A., K. A. Hobson, C. L. Gratto-Trevor, and A. W. Diamond. 1996. Conventional and isotopic determinations of shorebird diets at an inland stopover: The importance of invertebrates and *Potamogeton pectinatus* tubers. *Canadian Journal of Zoology* **74:**1057–1068.

Alisauskas, R. T., E. E. Klaas, K. A. Hobson, and C. D. Ankney. 1998. Stable-carbon isotopes support use of adventitious color to discern winter origins of lesser snow geese. *Journal of Field Ornithology* **69:**262–268.

Ambrose, S. H., and M. J. DeNiro. 1986. The isotopic ecology of East African mammals. *Oecologia* **69:**395–406.

Atkinson, P. W., A. J. Baker, R. M. Bevan, N. A. Clark, K. B. Cole, P. M. Gonzalez, J. Newton, L. J. Niles, and R. A. Robinson. 2005. Unravelling the migration and moult strategies of long-distance migrant using stable isotopes: Red Knot *Calidris canutus* movements in the Americas. *Ibis* **147:**738–749.

Battley, P. F., T. Piersma, M. W. Dietz, S. Tang, A. Dekinga, and K. Hulsman. 2000. Empirical evidence for differential organ reductions during trans-oceanic bird flight. *Proceedings of the Royal society of London, Series B* **267:**191–195.

Bearhop, S., D. R. Thompson, S. Waldron, I. C. Russell, G. Alexander, and R. W. Furness. 1999. Stable isotopes indicate the extent of freshwater feeding by cormorants *Phalacrocorax carbo* from inland fisheries in England. *Journal of Applied Ecology* **36:**75–84.

Bearhop, S., S. Waldron, S. C. Votier, and R. W. Furness. 2002. Factors that influence assimilation rates and fractionation of nitrogen and carbon stable isotopes in avian blood and feathers. *Physiological and Biochemical Zoology* **75:**451–458.

Bearhop, S., R. W. Furness, G. M. Hilton, S. C. Votier, and S. Waldron. 2003. A forensic approach to understanding diet and habitat use from stable isotope analysis of (avian) claw material. *Functional Ecology* **17:**270–275.

Bearhop, S., G. M. Hilton, S. C. Votier, and S. Waldron. 2004. Stable isotope ratios indicate that body condition in migrating passerines is influenced by winter habitat. *Proceedings of the Royal Society of London (Series B)* **271:**215–218.

Bearhop, S., W. Fiedler, R. W. Furness, S. C. Votier, S. Newton, J. Waldron, G. Bowen, P. Berthold, and K. Farnsworth. 2005. Assortative mating as a mechanism for rapid evolution of a migratory divide. *Science* **310:**502–504.

Bensch, S., G. Bengtsson, and S. Åkesson. 2006. Patterns of stable isotope signatures in willow warbler *Phylloscopus trochilus* feathers collected in Africa. *Journal of Avian Biology* **37:**323–330.

Best, P. B., and D. M. Schell. 1996. Stable isotopes in southern right whale (*Eubalaena australis*) baleen as indicators of seasonal movements, feeding and growth. *Marine Biology* **124**:483–494.

Birchall, J., T. C. O'Connel, T. H. E. Heaton, and R. E. M. Hedges. 2005. Hydrogen isotope ratios in animal body protein reflect trophic level. *Journal of Animal Ecology* **74**:877–881.

Bosley, K. L., D. A. Witting, R. C. Chambers, and S. C. Wainright. 2002. Estimating turnover rates of carbon and nitrogen in recently metamorphosed winter flounder *Psuedopleuronectes americana* with stable isotopes. *Marine Ecology Progress Series* **236**:233–240.

Boulet, M., H. L. Gibbs, and K. A. Hobson. 2006. Integrated analysis of genetic, stable isotope, and banding data reveal migratory connectivity and flyways in the northern Yellow Warbler (*Dendroica petechia*; *Aestiva* group). *Ornithological Monographs* **61**:29–78.

Bowen, G. J., L. I. Wassenaar, and K. A. Hobson. 2005. Application of stable hydrogen and oxygen isotopes to wildlife forensic investigations at global scales. *Oecologia* **143**:337–348.

Caccamise, D. F., L. M. Reed, P. M. Casteli, S. Wainsight, and T. C. Nichols. 2000. Distinguishing migratory and resident Canada geese using stable isotope analysis. *Journal of Wildlife Management* **64**:1084–1091.

Carleton, S. A., and C. Martínez del Rio. 2005. The effect of cold-induced increased metabolic rate on the ratio of ^{13}C and ^{15}N incorporation in house sparrows (*Passer domesticus*). *Oecologia* **144**:226–232.

Cerling, T. E., G. Bowen, J. R. Ehleringer, and M. Sponheimer. 2007. The reaction progress variable and isotope turnover in biological systems. Pages 163–171 *in* T. E. Dawson and R. T. W. Siegwolf (Eds.) *Stable Isotopes as Indicators of Ecological Change.* Academic Press, London.

Chamberlain, C. P., J. D. Blum, R. T. Holmes, X. Feng, T. W. Sherry, and G. R. Graves. 1997. The use of isotope tracers for identifying populations of migratory birds. *Oecologia* **109**:132–141.

Chamberlain, C. P., S. Bensch, X. Feng, S. Akesson, and T. Andersson. 2000. Stable isotopes examined across a migratory divide in Scandinavian willow Warblers (*Phylloscopus trochilus trochilus* and *Phylloscopus trochilus acredula*) reflect their African winter quarters. *Proceedings of the Royal Society (Series B)* **267**:43–48.

Clark, I. D., and P. Fritz. 1997. *Environmental Isotopes in Hydrogeology.* Lewis Publishers, New York.

Clark, R. G., K. A. Hobson, and L. I. Wassenaar. 2006. Geographic variation in the isotopic (δD, δ^{13}C, δ^{15}N, δ^{34}S) composition of feathers and claws from lesser scaup and northern pintail: Implications for studies of migratory connectivity. *Canadian Journal of Zoology* **84**:1395–1401.

Cormie, A. B., H. P. Schawarcz, and J. Gray. 1994. Relationship between the hydrogen and oxygen isotopes of deer bone and their use in the estimation of relative humidity. *Geochimica et Cosmochimica Acta* **60**:4161–4166.

Cryan, P. M., M. A. Bogan, R. O. Rye, G. P. Landis, and C. L. Kester. 2004. Stable hydrogen isotope analysis of bat hair as evidence for seasonal molt and long-distance migration. *Journal of Mammalogy* **85**:995–1001.

Dalerum, F., and A. Angerbjörn. 2005. Resolving temporal variation in vertebrate diets using naturally occurring stable isotopes. *Oecologia* **144**:647–658.

DeLong, J. P., T. D. Meehan, and R. B. Smith. 2005. Investigating fall movements of hatch-year Flammulated Owls (*Ottus flammeolus*) in central New Mexico using stable hydrogen isotopes. *Journal of Raptor Research* **39**:19–25.

Dockx, C., L. P. Brower, L. I. Wassenaar, and K. A. Hobson. 2004. Do North American monarch butterflies migrate to Cuba? Insights from combined isotope and chemical tracer techniques. *Ecological Applications* **14**:1106–1114.

Doucett, R. R., J. C. Marks, D. W. Blinn, M. Caron, and B. A. Hungate. 2007. Measuring terrestrial subsidies to aquatic foodwebs using stable-isotopes of hydrogen. *Ecology* **88**:1587–1592.

Drent, R. D. 2007. The timing of bird's breeding seasons: The Perrin's hypothesis revisited especially for migrants. *Ardea* **94**:305–322.

Dunn, E. H., K. A. Hobson, L. I. Wassenaar, D. Hussell, and M. L. Allen. 2006. Identification of summer origins of songbirds migrating through southern Canada in Autumn. *Avian Conservation and Ecology* 1:4 http://www.ace-eco.org/vol1/iss2/art4/.

Ehleringer, J. R., S. L. Phillips, W. S. F. Schuster, and D. R. Sandquist. 1991. Differential utilization of summer rains by desert plants. *Oecologia* **88**:430–434.

Evans-Ogden, L. J., K. A. Hobson, and D. B. Lank. 2004. Blood isotopic ($\delta^{13}C$ and $\delta^{15}N$) turnover and diet-tissue fractionation factors in captive Dunlin. *Auk* **121**:170–177.

Flemming, T. H., R. A. Nunez, and L. S. L. Sternberg. 1993. Seasonal changes in the diets of migrant and non-migrant necatrivorous bats as revealed by carbon stable isotope analysis. *Oecologia* **94**:72–75.

Gauthier, G., J. Bêty, and K. A. Hobson. 2003. Are greater snow geese capital breeders? New evidence from a stable isotope model. *Ecology* **84**:3250–3264.

Gómez-Diaz, E., and J. González-Solis. 2007. Geographic assignment of seabirds to their origin: Combining morphologic, genetic, and biogeochemical analyses. *Ecological Applications* **17**:1484–1498.

Gröcke, D. R., A. Schimmelmann, S. Elias, and R. F. Miller. 2006. Stable-hydrogen isotope ratios in beetle chitin: Preliminary European data and reinterpretation of North American data. *Quaternary Science Reviews* **25**:1850–1864.

Haramis, G. M., D. G. Jorde, S. A. Macko, and J. L. Walker. 2001. Stable isotope analysis of canvasback winter diet in Upper Chesapeake Bay. *Auk* **118**:1008–1017.

Harrington, R. R., B. P. Kennedy, C. P. Chamberlain, J. D. Blum, and C. L. Folt. 1998. [15]N enrichment in agricultural catchments: Field patterns and applications to tracking Atlantic salmon (*Salmo salar*). *Chemical Geology* **147**:281–294.

Heaton, T. H. E. 1987. The $^{15}N/^{14}N$ ratios of plants in South Africa and Namibia: Relationship to climate and coastal/saline environments. *Oecologia* **74**:236–246.

Hebert, C., and L. I. Wassenaar. 2001. Stable nitrogen isotopes in waterfowl feathers reflect agricultural land use in western Canada. *Environmental Science and Technology* **35**:3482–3487.

Hebert, C., and L. I. Wassenaar. 2005. Feather stable isotopes in western North American waterfowl: Spatial patterns, underlying factors, and management implications. *Wildlife Society Bulletin* **33**:92–102.

Hesslein, R. H., M. J. Capel, D. E. Fox, and K. A. Hallard. 1991. Stable isotopes of sulphur, carbon, and nitrogen as indicators of trophic level and fish migration in the lower Mackenzie River basin, Canada. *Canadian Journal of Fisheries and Aquatic Science* **48**:2258–2265.

Hesslein, R. H., K. A. Hallard, and P. Ramlal. 1993. Replacement of sulfur, carbon, and nitrogen in tissue of growing broad whitefish (*Coregonus nasus*) in response to a change in diet traced by $\delta^{34}S$, $\delta^{13}C$, and $\delta^{15}N$. *Canadian Journal of Fisheries and Aquatic Science* **50**:2071–2076.

Hobson, K. A. 1987. Use of stable-carbon isotope analysis to estimate marine and terrestrial protein content in gull diets. *Canadian Journal of Zoology* **65**:1210–1213.

Hobson, K. A. 1995. Reconstructing avian diets using stable-carbon and nitrogen isotope analysis of egg components: Patterns of isotopic fractionation and turnover. *Condor* **97**:752–762.

Hobson, K. A. 1999a. Tracing origins and migration of wildlife using stable isotopes: A review. *Oecologia* **120**:314–326.

Hobson, K. A. 1999b. Stable-carbon and nitrogen isotope ratios of songbird feathers grown in two terrestrial biomes: Implications for evaluating trophic relationships and breeding origins. *Condor* **101**:799–805.

Hobson, K. A. 2003. Making migratory connections with stable isotopes. Pages 379–391 *in* P. Berthold, E. Gwinner, and E. Sonnenschein (Eds.) *Avian Migration*. Springer-Verlag, Berlin Heidelberg, New York.

Hobson, K. A. 2005a. Stable isotopes and the determination of avian migratory connectivity and seasonal interactions. *Auk* **122**:1037–1048.

Hobson, K. A. 2005b. Using stable isotopes to trace long-distance dispersal in birds and other taxa. *Diversity and Distributions* **11**:157–164.

Hobson, K. A. 2006. Using stable isotopes to quantitatively track endogenous and exogenous nutrient allocations to eggs of birds that travel to breed. *Ardea* **94**:359–369.

Hobson, K. A. 2007. An isotopic exploration of the potential of avian tissues to track changes in terrestrial and marine ecosystems. Pages 129–144 *in* T. E. Dawson and R. T. W. Siegwolf (Eds.) *Stable Isotopes as Indicators of Ecological Change.* Academic Press, London.

Hobson, K. A., and E. Yohannes. 2007. Establishing elemental turnover in exercising birds using a wind tunnel: Implications for stable isotope tracking of migrants. *Canadian Journal of Zoology* **85**:703–708.

Hobson, K. A., and F. Bairlein. 2003. Isotopic discrimination and turnover in captive Garden Warblers (*Sylvia borin*): Implications for delineating dietary and migratory associations in wild passerines. *Canadian Journal of Zoology* **81**:1630–1635.

Hobson, K. A., and L. I. Wassenaar. 1997. Linking breeding and wintering grounds of Neotropical migrant songbirds using stable hydrogen isotopic analysis of feathers. *Oecologia* **109**:142–148.

Hobson, K. A., and L. I. Wassenaar. 2001. A stable isotope approach to delineating population structure in migratory wildlife in North America: An example using the loggerhead shrike. *Ecological Applications* **11**:1545–1553.

Hobson, K. A., and R. W. Clark. 1992. Assessing avian diets using stable isotopes. I: Turnover of carbon-13 in tissues. *Condor* **94**:181–188.

Hobson, K. A., R. T. Alisauskas, and R. G. Clark. 1993. Stable-nitrogen isotope enrichment in avian tissues due to fasting and nutritional stress: Implications for isotopic analysis of diet. *Condor* **95**:388–394.

Hobson, K. A., K. D. Hughes, and P. J. Ewins. 1997. Using stable-isotope analysis to identify endogenous and exogenous sources of nutrients in eggs of migratory birds: Applications to Great Lakes contaminants research. *Auk* **114**:467–478.

Hobson, K. A., L. Atwell, and L. I. Wassenaar. 1999a. Influence of drinking water and diet on the stable-hydrogen isotope ratios of animal tissues. *Proceedings of the National Academy of Science* **96**:8003–8006.

Hobson, K. A., L. I. Wassenaar, and O. R. Taylor. 1999b. Stable isotopes (δD and $\delta^{13}C$) are geographic indicators of natal origins of Monarch butterflies in eastern North America. *Oecologia* **120**:397–404.

Hobson, K. A., J. Sirois, and M. L. Gloutney. 2000a. Tracing nutrient allocations to reproduction using stable-isotopes: A preliminary investigation using the colonial waterbirds of Great Slave Lake. *Auk* **117**:760–774.

Hobson, K. A., R. B. Brua, W. L. Hohman, and L. I. Wassenaar. 2000b. Low frequency of "double molt" of remiges in ruddy ducks revealed by stable isotopes: Implications for tracking migratory waterfowl. *Auk* **117**:129–135.

Hobson, K. A., K. P. McFarland, L. I. Wassenaar, C. C. Rimmer, and J. E. Goetz. 2001. Linking breeding and wintering grounds of Bicknell's thrushes using stable isotope analyses of feathers. *Auk* **118**:16–23.

Hobson, K. A., L. I. Wassenaar, B. Milá, I. Lovette, C. Dingle, and T. B. Smith. 2003. Stable isotopes as indicators of altitudinal distributions and movements in an Ecuadorean hummingbird community. *Oecologia* **136**:302–308.

Hobson, K. A., L. I. Wassenaar, and E. Bayne. 2004a. Using isotopic variance to detect long-distance dispersal and philopatry in birds: An example with ovenbirds and American redstarts. *Condor* **106**:732–743.

Hobson, K. A., L. Atwell, L. I. Wassenaar, and T. Yerkes. 2004b. Estimating endogenous nutrient allocations to reproduction in redhead ducks: A dual isotope approach using δD and $\delta^{13}C$ measurements of female and egg tissues. *Functional Ecology* **18**:737–745.

Hobson, K. A., Y. Aubry, and L. I. Wassenaar. 2004c. Migratory connectivity in Bicknell's Thrush: Locating the missing populations using hydrogen isotopes. *Condor* **106**:905–909.

Hobson, K. A., G. Bowen, L. I. Wassenaar, Y. Ferrand, and H. Lormee. 2004d. Using stable hydrogen isotope measurements of feathers to infer geographical origins of migrating European birds. *Oecologia* **141**:477–488.

Hobson, K. A., J. E. Thompson, M. Evans, and S. Boyd. 2005. Tracing nutrient allocation to reproduction in Barrow's Goldeneye. *Journal of Wildlife Management* **69**:1221–1228.

Hobson, K. A., S. Van Wilgenburg, L. I. Wassenaar, H. Hands, W. Johnson, M. O'Melia, and P. Taylor. 2006. Using stable-hydrogen isotopes to delineate origins of Sandhill Cranes harvested in the Central Flyway of North America. *Waterbirds* **29**:137–147.

Hobson, K. A., R. J. F. Smith, and P. Sorensen. 2007a. Applications of stable isotope analysis to tracing nutrient sources to Hawaiian gobioid fish and other stream organisms. *Bishop Museum Bulletin in Cultural and Environmental Studies* **3**:99–111.

Hobson, K. A., S. Van Wilgenburg, L. I. Wassenaar, F. Moore, and J. Farrington. 2007b. Estimating origins of three species of Neotropical migrants at a Gulf coast stopover site: Combining stable isotope and GIS tools. *Condor* **109**:256–267.

Hobson, K. A., R. G. Clark, L. I. Wassenaar, and S. Van Wilgenburg. Natal origins of harvested Lesser Scaup (*Athya affinis*) in North America: Making connections to source populations with deuterium analyses of feathers. *Ecological Applications*, in review.

Ibargüen, S. B. 2004. *Population Connectivity: Combining Methods for Estimating Avian Dispersal and Migratory Linkages*. Ph.D. dissertation, Ohio State University. [Online.] Available at http://www.ohiolink.edu.proxy.lib.ohio-state.edu/etd/view.cgi?acc_num=osu1079979416.

Kelly, J. F. 2000. Stable isotopes of carbon and nitrogen in the study of avian and mammalian trophic ecology. *Canadian Journal of Zoology* **78**:1–27.

Kelly, J. F., V. Atudorei, Z. D. Sharp, and D. M. Finch. 2002. Insights into Wilson's Warbler migration from analyses of hydrogen stable-isotope ratios. *Oecologia* **130**:216–221.

Kennedy, B. P., C. L. Folt, J. D. Blum, and C. P. Chamberlain. 1997. Natural isotope markers in salmon. *Nature* **387**:766.

Killingly, J. S. 1980. Migrations of California gray whales tracked by oxyen-18 variations in their epizoic barnacles. *Science* **207**:759–760.

Killingly, J. S., and M. Lutcavage. 1983. Loggerhead turtle movements reconstructed from 18O and 13C profiles from commensal barnacle shells. *Estuarine and Coastal Shelf Science* **16**:345–349.

Klaassen, M., A. Lindstrom, H. Meltofte, and T. Piersma. 2001. Arctic waders are not capital breeders. *Nature* **413**:794.

Koch, P. L., J. Heisinger, C. Moss, R. W. Carlson, M. L. Fogel, and A. K. Behrensmeyer. 1995. Isotopic tracking of change in diet and habitat use in African elephants. *Science* **267**:1340–1343.

Krouse, H. R., J. W. B. Stewart, and V. A. Grinenko. 1991. Pedosphere and biosphere. Pages 267–306 *in* H. R. Krouse and V. A. Grinenko (Eds.) *Stable Isotopes: Natural and Anthropogenic Sulphur in the Environment.* John Wiley and Sons, Toronto.

Langin, K. M., M. W. Reudink, P. R. Marra, D. R. Norris, T. K. Kyser, and L. M. Ratcliffe. 2007. Hydrogen isotopic variation in migratory bird tissues of known origin: Implications for geographic assignment. *Oecologia* **152**:449–457.

Losey, J. E., L. S. Rayor, and M. E. Carter. 1999. Transgenic pollen harms monarch larvae. *Nature* **399**:214–214.

Lott, C. A., and J. P. Smith. 2006. A GIS approach to estimating the origins of migratory raptors in North America using hydrogen stable isotope ratios in feathers. *Auk* **118**:16–23.

Lott, C. A., T. D. Meehan, and J. A. Heath. 2003. Estimating the latitudinal origins of migratory birds using hydrogen and sulfur stable isotopes in feathers: Influence of marine prey base. *Oecologia* **134**:505–510.

Marra, P. P., K. A. Hobson, and R. T. Holmes. 1998. Linking winter and summer events in a migratory bird using stable carbon isotopes. *Science* **282**:1884–1886.

Marquiss, M., K. A. Hobson, and I. Newton, Stable isotope evidence for regionalised provenances of Common Crossbill *Loxia curvirostra* invasions into Western Europe. *Journal of Avian Biology*, in press.

Mazerolle, D., and K. A. Hobson. 2005. Estimating origins of short-distance migrant songbirds in North America: Contrasting inferences from hydrogen isotope measurements of feathers, claws, and blood. *Condor* **107**:280–288.

Mazerolle, D., K. A. Hobson, and L. I. Wassenaar. 2005. Stable isotope and band-encounter analyses delineate migratory patterns and catchment areas of white-throated sparrows at a migration monitoring station. *Oecologia* **144**:541–549.

McKechnie, A. E., B. O. Wolf, and C. Martinez del Rio. 2004. Deuterium stable isotope ratios as tracers of water resource use: An experimental test with rock dove. *Oecologia* **140**:191–200.

Meehan, T. D., C. A. Lott, Z. D. Sharp, R. B. Smith, R. N. Rosenfield, A. C. Stewart, and R. K. Murphy. 2001. Using hydrogen isotope geochemistry to estimate the natal latitudes of immature cooper's hawks migrating through the Florida Keys. *Condor* **103**:11–20.

Meehan, T. D., R. N. Rosenfield, V. N. Atudorei, J. Bielefeldt, L. J. Rosenfield, A. C. Stewart, W. E. Stout, and M. A. Bozek. 2003. Variation in hydrogen stable-isotope ratios between adult and nestling Cooper's hawk. *Condor* **105**:567–572.

Meehan, T. D., J. T. Giermakowski, and P. M. Cryan. 2004. A GIS-based model of stable hydrogen isotope ratios in North American growing-season precipitation for use in animal movement studies. *Isotopes in Environmental and Health Studies* **40**:291–300.

Mehl, K. R., R. T. Alisauskas, K. A. Hobson, and F. R. Merkel. 2005. Linking breeding and wintering grounds of king eiders: Making use of polar isotopic gradients. *Journal of Wildlife Management* **69**:1297–1304.

Meyer-Rochow, V. B., I. Cook, and H. Hendy. 1992. How to obtain clues from the otoliths of an adult fish about the aquatic environment it has been in as a larvae. *Comparative Biochemistry and Physiology* **103A**:333–335.

Mizutani, H., M. Fukuda, Y. Kabaya, and E. Wada. 1990. Carbon isotope ratio of feathers reveals feeding behavior of cormorants. *Auk* **107**:400–403.

Møller, A. P., and K. A. Hobson. 2004. Heterogeneity in stable isotope profiles predicts coexistence of two populations of barn swallows *Hirundo rustica* differing in morphology and reproductive performance. *Proceedings of the Royal Society of London* **271**:1355–1362.

Møller, A. P., K. A. Hobson, T. A. Mousseau, and A. M. Peklo. 2006. Chernobyl as a population sink for barn swallows: Tracking dispersal using stable isotope profiles. *Ecological Applications* **16**:1696–1705.

Morrison, R. I. G., and K. A. Hobson. 2004. Use of body stores in shorebirds after arrival on High Arctic breeding grounds. *Auk* **121**:333–344.

Nelson, C. S., T. G. Northcote, and C. H. Hendy. 1989. Potential use of oxygen and carbon isotopic composition of otoliths to identify migratory and non-migratory stocks of the New Zealand common smelt: A pilot study. *New Zealand Journal of Marine and Freshwater Research* **23**:337–344.

Newton, I., K. A. Hobson, A. D. Fox, and M. Marquiss. 2006. An investigation into the provenance of Northern Bullfinches *Pyrrhula p. pyrrhula* found in winter in Scotland and Denmark. *Journal of Avian Biology* **37**:431–435.

Norris, D. R., P. R. Marra, T. K. Kyser, T. W. Sherry, and L. M. Ratcliffe. 2004a. Tropical winter habitat limits reproductive success on the temperate breeding grounds in a migratory bird. *Proceedings of the Royal Society of London (Series B)* **271**:59–64.

Norris, D. R., P. R. Marra, R. Montgomerie, T. K. Kyser, and L. M. Ratcliffe. 2004b. Reproductive effort, molting latitude, and feather color in a migratory songbird. *Science* **306**:2249–2250.

Norris, D. R., P. R. Marra, T. K. Kyser, and L. M. Ratcliffe. 2005. Tracking habitat use in a long distance migratory bird, the American Redstart *Setophaga ruticilla*, using stable isotope ratios in cellular blood. *Journal of Avian Biology* **36**:164–170.

Norris, D. R., P. R. Marra, G. J. Bowen, L. M. Ratcliffe, J. A. Royle, and T. K. Kyser. 2006. Migratory connectivity of a widely distributed songbird, the American Redstart (*Setophaga ruticilla*). *Ornithological Monographs* **61**:14–28.

Norris, D. R., P. R. Marra, K. T. Kyser, L. M. Ratcliffe, and R. Montgomerie. 2007. Continent wide variation in feather color of a migratory songbird in relation to body condition and molting locality. *Biology Letters* **3**:16–19.

Ostrom, P. H., M. Colunga-Garcia, and S. H. Gage. 1997. Establishing pathways of energy flow for insect predators using stable isotope ratios: Field and laboratory evidence. *Oecologia* **109**:108–113.

Pain, D. J., R. E. Green, B. Giessing, A. Kozulin, A. Poluda, U. Ottosson, M. Flade, and G. M. Hilton. 2004. Using stable isotopes to investigate migratory connectivity of the globally threatened aquatic warbler *Acrocephalus paludicola*. *Oecologia* **138**:168–174.

Paxton, K. L., C. Van Riper III, T. C. Theimer, and E. H. Paxton. 2007. Spatial and temporal migration patterns of Wilson's Warbler (*Wilsonia pusilla*) in the southwest as revealed by stable isotopes. *Auk* **124**:162–175.

Pearson, S. F., D. J. Levey, C. H. Greenberg, and C. Martinez del Rio. 2003. Effects of elemental composition on the incorporation of dietary nitrogen and carbon isotopic signatures in an omnivorous songbird. *Oecologia* **135**:516–523.

Perez, G., and K. A. Hobson. 2006. Isotopic evaluation of interrupted molt of northern breeding populations of the loggerhead shrike. *Condor* **108**:877–886.

Perez, G., and K. A. Hobson. 2007. Feather deuterium measurements reveal origins of migratory western Loggerhead Shrikes (*Lanius ludovicianus excubitorides*) wintering in Mexico. *Diversity and Distributions* **13**:166–171.

Phillips, D. L., and P. M. Eldridge. 2006. Estimating the timing of diet shifts using stable isotopes. *Oecologia* **147**:195–203.

Piersma, T., G. A. Gudmundsson, and K. Lilliendahl. 1999. Rapid changes in the size of different functional organ and muscle groups during refuelling and a long-distance migrating shorebird. *Physiological and Biochemical Zoology* **72**:405–415.

Podlesak, D. W., S. R. McWilliams, and K. A. Hatch. 2005. Stable isotopes in breath, blood, feces and feathers can indicate intra-individual changes in the diet of migratory songbirds. *Oecologia* **142**:501–510.

Post, D. M. 2002. Using stable isotopes to estimate trophic position: Models, methods, and assumptions. *Ecology* **83**:703–718.

Powell, L. A., and K. A. Hobson. 2006. Enriched feather hydrogen isotope values for Wood Thrushes sampled in Georgia, USA, during the breeding season: Implications for quantifying dispersal. *Canadian Journal of Zoology* **84**:1331–1338.

Robbins, C. T., L. A. Felicetti, and M. Sponheimer. 2005. The effect of dietary protein quality on nitrogen isotope discrimination in mammals and birds. *Oecologia* **144**:534–540.

Rocque, D. A., M. Ben-David, R. P. Barry, and K. Winker. 2006. Assigning birds to wintering and breeding grounds using stable isotopes: Lessons from two feather generations among three intercontinental migrants. *Journal of Ornithology* **147**:395–404.

Rubenstein, D. R., and K. A. Hobson. 2004. From birds to butterflies: Animal movement patterns and stable isotopes. *Trends in Ecology and Evolution* **19**:256–263.

Rubenstein, D. R., C. P. Chamberlain, R. T. Holmes, M. P. Ayres, J. R. Waldbauer, G. R. Graves, and N. C. Tuross. 2002. Linking breeding and wintering ranges of a migratory songbird using stable isotopes. *Science* **295**:1062–1065.

Schell, D. M., S. M. Saupe, and N. Haubenstock. 1989. Bowhead whale (*Balaena mysticetus*) growth and feeding as estimated by $\delta^{13}C$ techniques. *Marine Biology* **103**:433–443.

Sealy, J. C., N. J. van der Merwe, J. A. Lee Thorp, and J. L. Lanham. 1987. Nitrogen isotopic ecology in southern Africa: Implications for environmental and dietary tracing. *Geochimica et Cosmochimica Acta* **51**:2707–2717.

Sheppard, S. M. F., R. L. Neilsen, and H. P. Taylor. 1969. Oxygen and hydrogen isotope ratios of clay minerals from porphyry copper deposits. *Economic Geology* **64**:755–777.

Smith, A. D., and A. M. Dufty. 2005. Variation in the stable-hydrogen isotope composition of Northern Goshawk feathers: Relevance to the study of migratory origins. *Condor* **107**:547–558.

Szymanski, M., A. Afton, and K. A. Hobson. 2006. Use of stable isotope methodology to determine natal origin of hatch-year mallards shot during fall in Minnesota. *Journal of Wildlife Management* **71**:1317–1324.

Taylor, H. P., Jr. 1974. An application of oxygen and hydrogen isotope studies to problems of hydrothermal alteration and ore deposition. *Economic Geology* **69**:843–883.

Tieszen, L. L., T. W. Boutton, K. G. Tesdahl, and N. A. Slade. 1983. Fractionation and turnover of stable carbon isotopes in animal tissues: Implications for $\delta^{13}C$ analysis of diet. *Oecologia* **57**:32–37.

Tietje, W. D., and J. G. Teer. 1988. Winter body condition of Northern Shovelers on freshwater and saline habitats. Pages 353–377 *in* D. J. Batt, R. H. Chaebreck, L. H. Fredrickson, and D. G. Raveling (Eds.) *Waterfowl in Winter*. University of Minnesota Press, Minneapolis.

Trueman, C. N., and A. Moore. 2007. Use of the stable isotope composition of fish scales for monitoring aquatic ecosystems. Pages 145–161 *in* T. E. Dawson and R. T. W. Siegwolf (Eds.) *Stable Isotopes as Indicators of Ecological Change*. Academic Press, London.

Trust, B. 1993. *Stable Carbon and Sulfur Isotopic Ratios in Migrating Redhead Ducks* (Aythya americana) *and Their Breeding and Wintering Habitats*. Ph D thesis, University of Texas, Austin.

Vanderklift, M. A., and S. Ponsard. 2003. Sources of variation in consumer-diet $\delta^{15}N$ enrichment: A meta analysis. *Oecologia* **136**:169–182.

Van der Merwe, N. J., J. A. Lee Thorp, J. F. Thackeray, A. Hall-Martin, F. J. Kruger, H. Coertzees, R. H. V. Bell, and M. Lindeque. 1990. Source-area determination of elephant ivory by isotopic analysis. *Nature* **346**:744–746.

Vogel, J. C., B. Eglington, and J. M. Auret. 1990. Isotopic fingerprints in elephant bone and ivory. *Nature* **346**:747–749.

Walpole, R. E., and R. H. Meyers. 1993. *Probability and Statistics for Engineers and Scientists*. 5th edn. Prentice Hall, Englewood Cliffs, New Jersey.

Wassenaar, L., and K. A. Hobson. 1998. Natal origins of migratory Monarch butterflies at wintering colonies in Mexico: New isotopic evidence. *Proceedings of the National Academy of Sciences* **95**:15436–15439.

Wassenaar, L. I., and K. A. Hobson. 2000. Stable-carbon and hydrogen isotope ratios reveal breeding origins of red-winged blackbirds. *Ecological Applications* **10**:911–916.

Wassenaar, L. I., and K. A. Hobson. 2001. A stable-isotope approach to delineate geographical catchment areas of avian migration monitoring stations in North America. *Environmental Science and Technology* **35**:1845–1850.

Wolf, B., and C. Martinez del Rio. 2000. Use of saguaro fruit by white-winged doves: Isotopic evidence of a tight ecological association. *Oecologia* **124**:536–543.

Wunder, M. B. 2007. *Geographic Structure and Dynamics in Mountain Plover*. Ph.D. dissertation. Colorado State University, Fort Collins, Colorado.

Wunder, M. B., C. L. Kester, F. L. Knopf, and R. O. Rye. 2005. A test of geographic assignment using isotope tracers in feathers of known origin. *Oecologia* **144**:607–617.

Yapp, C. J., and S. Epstein. 1982. A re-examination of cellulose carbon-bound hydrogen D measurements and some factors affecting plant-water D/H relationships. *Geochimica et Cosmochimica Acta* **46**:955–965.

Yohannes, E., K. A. Hobson, D. Pearson, L. I. Wassenaar, and H. Biebach. 2005. Stable isotope analyses of feathers help identify autumn stopover sites of three long-distance migrants in northeastern Africa. *Journal of Avian Biology* **36**:235–241.

Yohannes, E., K. A. Hobson, and D. J. Pearson. 2007. Feather stable-isotope profiles reveal stopover habitat selection and site fidelity in nine migratory species moving through sub-Saharan Africa. *Journal of Avian Biology* **38**:347–355.

CHAPTER 4

Isotope Landscapes for Terrestrial Migration Research

Gabriel J. Bowen* and Jason B. West[†]

*Earth and Atmospheric Sciences Department, Purdue University
[†]Department of Biology, University of Utah

Contents

Tracking Animal Migration with Stable Isotopes
K. A. Hobson and L. I. Wassenaar (Editors)
ISSN 1936-7961, DOI: 10.1016/S1936-7961(07)00004-8

I. INTRODUCTION

Isotope tracking of migratory terrestrial animals (*e.g.*, birds, bats, and insects) relies on the assimilation and fixation of intrinsic isotopic markers from the environment into animal body tissue. The power of the isotopic markers relates to the extent and pattern of spatial isotope ratio variations in the environmental substrates from which they are assimilated (primarily food, water, and air). This chapter introduces these patterns of variation for most of the commonly applied or applicable stable isotope systems, and describes methods by which the spatial landscapes of environmental isotopic variation ("isoscapes") are modeled and predicted at scales relative to the study of migratory behavior and ecology. Examples of isoscapes for some isotopic systems are presented along with discussion of challenges and cautionary notes related to the creation and interpretation of isoscapes. A discussion of opportunities and future directions in isoscape modeling is offered. The goal of this chapter is to familiarize the researcher with isoscape data products and the range of current and potential products for use in migration applications. Also described are the principles and methodology underlying the development of these products and relevant to their informed use.

Maps of some isotopic landscapes have been available since the early 1980s, when workers affiliated with the International Atomic Energy Agency (IAEA) and World Meteorological Organization's Global Network for Isotopes in Precipitation (GNIP) compiled and contoured mean annual precipitation isotope ratio data to produce a map with near-global coverage (Yurtsever and Gat 1981). The GNIP data set has remained a prime example of a spatial isotope monitoring network, providing a data set that has motivated the development of improved isoscapes for H and O in water (Birks *et al.* 2002, Bowen and Wilkinson 2002, Bowen and Revenaugh 2003, Meehan *et al.* 2004, Bowen *et al.* 2005, Bowen *et al.* 2007) and, more recently, for plants (West *et al.* in review).

Isoscapes of ecosystem-scale carbon isotope ratios were found to be relevant to the study of the global carbon cycle using $\delta^{13}C$ and have been in development since the early 1990s (Lloyd and Farquhar 1994, Still *et al.* 2003, Suits *et al.* 2005). The development of isoscapes for other isotopic systems is very recent, and preliminary attempts to represent plant and soil nitrogen isoscapes (Amundson *et al.* 2003) and Sr isoscapes (Beard and Johnson 2000) have appeared in the literature in the last few years.

Research on the spatial patterning of isotopes in the environment and their modeling and depiction is an active field, and although some data products are well documented and publicly available through websites such as http://www.waterisotopes.org, the scope of these products is currently limited. Given that much of the data, theory, and software enabling isoscapes modeling are freely available, a secondary goal of this chapter is to introduce many of the fundamental considerations underlying isoscapes modeling in hope of encouraging additional researchers to participate in the future research and development of isoscapes relevant to migration research, in particular to tackle such questions as connectivity, movement across landscapes, and intra- and interspecific habitat use at various spatiotemporal resolutions.

II. PROCESS

Mapping isotopic variations across space is accomplished through the identification, simplification, and modeling of the processes that lead to isotopic variations at the landscape level. Although a wide range of physical and chemical processes can produce isotopic discrimination and contribute to observed isotope distributions, a relatively small subset of processes dominates landscape-level variability. We begin by highlighting and reviewing these processes as they relate to isoscape modeling for terrestrial migration studies.

A. Rayleigh Distillation and Precipitation $\delta^2 H$ and $\delta^{18} O$

Global spatial variability in the isotopic composition of meteoric precipitation (rainfall) was first reported by Dansgaard (1954, see also Craig 1961, Dansgaard 1964), making it one of the longest-studied examples of landscape-level isotopic variation. The profound variation in $\delta^2 H$ and $\delta^{18} O$ values measured for precipitation samples collected worldwide (Rozanski *et al.* 1993) can largely be attributed to a single phenomenon: the progressive drying of air masses as they lose moisture in the form of precipitation. The phase change reaction that leads to the formation of water droplets (or ice crystals) in clouds proceeds with an equilibrium isotope effect α_e through which water molecules containing heavy isotopic species are preferentially incorporated in the liquid or solid phase during condensation (note that α_e differs for condensation to liquid vs solid phase and for $^2 H$- and $^{18} O$-bearing water molecules). Having grown to a size at which they fall from the convecting cloud mass, rain droplets or snow crystals are effectively removed from the cloud system, taking with them a disproportionate concentration of the "heavy" isotopic species and leaving the cloud vapor incrementally depleted in $^2 H$ and $^{18} O$. As this process proceeds, the $\delta^2 H$ and $\delta^{18} O$ values of the cloud vapor become progressively lower according to the Rayleigh equation (given in terms of ratios):

$$R = R_0 f^{(\alpha-1)} \tag{4.1}$$

where R is the isotope ratio of cloud vapor at any point in time, R_0 is the initial isotope ratio of the air mass, and f is the fraction of vapor remaining. For the residual vapor produced through the condensation process, values of α are less than 1, giving a progressive decrease in vapor isotope ratios (and those of newly formed precipitation) as an air mass dries.

B. Hydrological Mixing, Evaporation, and Surface Water $\delta^2 H$ and $\delta^{18} O$

The primary postprecipitation processes affecting the stable isotopic composition of continental waters are hydrological mixing of waters having different isotopic compositions and reevaporation from the land surface. Large data sets of H and O isotope ratios of river and tap water clearly show the impact of both processes on spatial variation in environmental water isotopic compositions (Kendall and Coplen 2001, Bowen *et al.* 2007). Mixing represents a linear process, weighed according to the volumetric concentration of the different sources. Mixing is a particularly important process where waters are present that have very different isotopic compositions because they fell at different times or places (*e.g.*, mountain snowmelt and summertime basinal rain in mountainous regions) or have different postprecipitation histories (*e.g.*, thermal spring water and surface water).

Reevaporation of water can occur at many points in the postprecipitation history of all continental waters, including from leaf surfaces that intercept falling rain, soils, rivers, lakes, or reservoirs. Isotope fractionation during evaporation does not follow a simple equilibrium model but involves a dynamic balance of the phase change reaction and bidirectional diffusive transport between a boundary layer adjacent to the liquid surface and the free atmosphere. These processes were synthesized as the "Craig–Gordon model" (Craig and Gordon 1965):

$$\frac{d\delta}{d\ln f} = \frac{h(\delta - \delta_a) - (\delta + 1)(\Delta\varepsilon + \varepsilon/\alpha)}{1 - h + \Delta\varepsilon} \varepsilon_t = \varepsilon_e + \Delta\varepsilon_k, \tag{4.2}$$

where δ and δ_a are the isotopic compositions of the evaporating water body and atmosphere, f is the remaining fraction of liquid, h is atmospheric humidity, and ε, α, and $\Delta\varepsilon$ are the isotope effect and fractionation factor for equilibrium evaporation, and the kinetic isotope effect of evaporation, respectively. Under low-humidity conditions, evaporation can be approximated as distillation process following Eq. (4.1). The net isotope effect of evaporation is to increase the δ^2H and $\delta^{18}O$ values of residual surface water, with the magnitude of change scaling with the extent of evaporation and affected by h and other parameters that determine the evaporative fractionation.

C. Plant Water and Organic H and O Isotopes

The rainfall isotopic patterns discussed in Section II.B are intimately connected to the base of all terrestrial food webs through the linkage of water and primary productivity (plants). To a first approximation, the isotopic composition of the water in plants may be simplified to two pools: unfractionated water that essentially matches the plant source water isotopic composition (*e.g.*, xylem water derived from soil water) and ^2H- and ^{18}O-enriched leaf water. Leaf water isotope ratios increase in response to evaporation in a manner analogous to open water bodies, with liquid to gas phase changes occurring inside the leaf, diffusion of vapor through stomatal openings and the leaf boundary layer, and isotopic exchange of leaf water with atmospheric vapor. These isotope effects have been described in several models of leaf water enrichment, derived principally from the Craig–Gordon model. The general equation for steady state leaf water isotope ratios has been written as:

$$R_e = \alpha \left[\alpha_k R_S \left(\frac{e_i - e_s}{e_i} \right) + \alpha_{kb} R_S \left(\frac{e_s - e_a}{e_i} \right) + R_A \left(\frac{e_a}{e_i} \right) \right] \qquad (4.3)$$

where R_e is the isotope ratio of evaporated leaf water, R_S is the isotope ratio of the source water, R_A is the isotope ratio of the atmospheric water vapor, e_i is internal leaf vapor pressure, e_s is the leaf surface vapor pressure, and e_a is atmospheric vapor pressure (Flanagan *et al.* 1991). Other models incorporate additional complexities found to be important, including nonsteady state dynamics (Dongmann *et al.* 1974, Farquhar and Cernusak 2005) and within-leaf heterogeneity, including back-diffusion of heavy isotopes in liquid leaf water (Yakir *et al.* 1994, Helliker and Ehleringer 2000, Péclet effect; Farquhar and Cernusak 2005).

Leaf water isotopic composition is important both because leaf water is a potential source of animal body water (*e.g.*, Murphy *et al.* 2007) and because the isotopic composition of plant organic molecules is dependent on the isotopic composition of the water at sites of photosynthesis in leaves, as well as the isotopic effects of photosynthesis itself and other biochemical transformations during plant metabolism (Roden *et al.* 2000). These isotope fractionations can be relatively large. For example, observed fractionation factors for the formation of cellulose are $\varepsilon = 27‰$ for oxygen isotopes and $\varepsilon = 158‰$ for hydrogen isotopes (Sternberg and Deniro 1983, Yakir *et al.* 1990, Luo and Sternberg 1992).

D. Gas Exchange, Photosynthetic Pathway, and Plant C Isotopes

The carbon isotope ratios of plant tissues are dependent on the $\delta^{13}C$ of atmospheric CO_2 and the isotope fractionation events that occur during carbon fixation, including diffusion, dissolution, and hydration of CO_2, and the fixation reactions themselves (Park and Epstein 1960, Osmond *et al.* 1973, Farquhar *et al.* 1989, O'Leary *et al.* 1992, Von Caemmerer 1992, Ehleringer and Monson 1993). There are three primary pathways of carbon fixation in plants: C_3, C_4, and CAM, names that refer to the number of carbon atoms in the first stable product of photosynthesis (C_3 and C_4) or the nocturnal buildup of malic acid (CAM). The C_3 pathway utilizes ribulose-1,5-bisphosphate carboxylase/oxygenase (Rubisco) as a catalyst to form a three-carbon molecule from atmospheric CO_2 during the day. The C_4 pathway is primarily found in grasses and utilizes phosphoenolpyruvate (PEP) carboxylase as a catalyst to fix HCO_3^- (atmospheric CO_2 hydrated in a reaction catalyzed by carbonic anhydrase) initially to a four-carbon molecule that is then the source of CO_2 for C_3 photosynthesis. In C_4 plants, the C_3 photosynthesis phase is generally spatially isolated from the atmosphere and receives high concentrations of CO_2 from the PEP carboxylase system. CAM plants also utilize PEP carboxylase to catalyze the first step, but rather than isolate Rubisco spatially as occurs in C_4 plants, these plants do the initial step of fixing CO_2 at night when the vapor pressure deficit is lower. Rubisco strongly discriminates against $^{13}CO_2$ ($\sim$30‰ O'Leary 1981) giving C_3 plants strongly ^{13}C-depleted carbon isotope ratios. PEP carboxylase, on the other hand, shows much lower discrimination ($\sim$2‰) that results in C_4 plants having higher $\delta^{13}C$ values (CAM plants show a wide range of values; Ehleringer *et al.* 1986).

The differences in isotopic discrimination between these two enzymes are large enough that C_3 and C_4 plants have nearly completely nonoverlapping carbon isotope distributions (Ehleringer *et al.* 1986). This was observed very early in the study of stable isotope ratios in plants (Wickman 1952, Craig 1953), and has been utilized extensively to understand plant–animal interactions, both in modern environments and in paleoenvironments (*e.g.*, Van Der Merwe *et al.* 1988, Cerling *et al.* 2003).

Stomatal resistance has an important effect on the carbon isotope ratios of C_3 plants. The number and aperture of stomates control the CO_2 diffusion resistance between the atmosphere and the substomatal cavities, and thus determines the relative openness of the isotopic system with respect to the source CO_2. As stomates close, the supply of CO_2 to the substomatal cavity, as well as its diffusion back out of the leaf, slows. This slowing of CO_2 exchange with the atmosphere results in a relative ^{13}C enrichment of the products of photosynthesis because of a progressive ^{13}C enrichment of the internal leaf CO_2 pool and a proportional increase in the flux of $^{13}CO_2$ into the reactions of photosynthesis. Because of this effect of stomatal aperture, climate exerts a dominant influence on the carbon isotopic signature of C_3 plant tissues. With greater water limitation, stomates tend to close, increasing the $\delta^{13}C$ of the resulting sugars and ultimately other organic materials in plants (Hemming *et al.* 2005).

E. Nitrogen Isotopes in Soils and Plants

Plant $\delta^{15}N$ is determined by the isotopic composition of the plant N source and the isotope fractionations associated with plant N uptake and metabolism. Because there are multiple, competing reactions in soils, many of which have strong isotopic effects, a general theory describing soil $\delta^{15}N$ has not yet emerged (Evans 2001, Robinson 2001). Primarily because of this, plant nitrogen isotopic composition has been generally interpreted as an integrator of the effects of important processes that can inform questions related to changes in N cycling and sources, but that needs to be interpreted cautiously and in primarily site-specific ways (Evans 2001, Robinson 2001). Large-scale comparisons have shown correlations between both the $\delta^{15}N$ of plants and soils and climate variables such as mean annual temperature and precipitation (Austin and Sala 1999, Handley *et al.* 1999, Amundson *et al.* 2003, Ometto *et al.* 2006). Although much of the global observed variability remains unexplained by these simple models, they do suggest dominant controls on soil–plant $\delta^{15}N$ based on preferential losses of ^{15}N from systems (cf. Houlton *et al.* 2006). Clearly, inputs are an important control on plant and soil $\delta^{15}N$ also because N fixation inputs are near 0‰, but can vary from -3‰ to 1‰ (West *et al.* 2005) and nitrogen deposition can vary considerably, depending on the nitrogen source (Pardo *et al.* 2006, Widory 2007). More recent work suggests the tantalizing possibility that remotely sensed reflectance data might yield information on vegetation $\delta^{15}N$ (Wang *et al.* 2007).

F. Ecosystem Sr Isotopes

Isotopic variation in Sr, unlike that in the other light isotope systems discussed here, is driven by the continuous production of one of the common isotopes, ^{87}Sr, which forms during the β^- decay of ^{87}Rb. Both the radiogenic isotope ^{87}Sr and the more common, nonradiogenic ^{86}Sr are stable, and differences in the relative abundance of these isotopes due to the level of production of the radiogenic isotope are tracers of Sr source. Because the half-life of ^{87}Rb is quite long ($\sim$48.8 billion years), production of ^{87}Sr occurs over geological timescales and much of the potential Sr isotope variability in and among ecosystems is related to the type and history of the rocks from which their Sr is derived. Variation in the ^{87}Sr/^{86}Sr of rocks occurs for a number of different reasons, including variation in the initial Sr isotope ratio of igneous rocks and carbonate rocks at the time they solidify from magma, variation in the ^{87}Rb content of different rocks, and variation in the age of rocks (Faure and Powell 1972).

When considering variation in ecosystem Sr isotope ratios that might be sampled by a migrating animal, however, isotopic variability of rock Sr only represents the base layer of variability at the landscape level [*e.g.,* see review by Capo *et al.* (1998)]. Rock Sr is released to ecosystems by weathering, and in areas where multiple rock sources are available, Sr isotope ratios within ecosystems can be heavily biased toward those of Sr sourced from individual rock types. In many cases, ecosystem ^{87}Sr/^{86}Sr is largely determined by those of carbonate rocks that weather rapidly and have total Sr concentrations approximately two to six times higher than most other rocks (Faure and Powell 1972). Sources of Sr entering groundwater, surface water, and soils can often differ, particularly in areas where windblown dust contributes a significant fraction of Sr to soils (*e.g.,* Quade *et al.* 1995, Kennedy *et al.* 1998) or where bedrock geology is particularly heterogeneous, meaning that multiple, isotopic distinct pools of Sr may be available to plants and animals living in proximity. Nonetheless, spatial variability in ^{87}Sr/^{86}Sr, primarily because of major variation in rock type distributions over large spatial scales, has been shown to propagate into ecosystems and represent a useful tracer for migratory ecology and paleoecology (*e.g.,* Chamberlain *et al.* 1997, Hoppe *et al.* 1999).

III. PATTERN

The processes discussed in the previous sections produce spatial isotope variations in the environment, but in order to understand the existence of systematic, predictable, isotopic variability that is useful for terrestrial migration research, we must consider the organization of these processes with respect to space. Indeed, the fact that many isotopically discriminating processes are spatially patterned underlies the isoscapes concept and the correlative application of stable isotope tracking in migration research. The creation of isoscapes for migration research requires that these patterns be mapped in space, usually through the mathematical transformation of maps depicting correlates or environmental drivers of isotopic variation. We will first review the origin of these patterns, considering how they relate to geological, physiographic, climatological, and biological (collectively, "environmental") drivers, and then introduce a two-endmember classification system for spatial patterning of isotopic variation.

A. Spatial Organization and Isotope Patterning

Stable isotope variations in the environment reflect the spatial patterning of environmental factors in three primary ways. First, geographic position can determine the elemental sources available for incorporation in biological substrates, and thus impact the isotopic composition of ecologically relevant materials. Second, where isotopes are fractionated by environmental processes, the magnitude of fractionation at a given site will respond to local environmental conditions. Third, for systems in which large-scale geographic transport is important, the spatial organization of transport processes (*e.g.,* atmospheric circulation, runoff, groundwater flow) will determine both the isotopic source and the integrated history of fractionating processes, affecting material in the local environment. Each of these modes relating isotopic variation to spatially varying environmental parameters can be illustrated through examples from the systems introduced earlier.

Spatial variation in the isotopic composition of elemental sources is important in all systems where these sources exhibit landscape-level isotopic heterogeneity, and in many cases may be the primary determinant of the isotopic composition of biological systems. Strontium isotopes, which are not measurably fractionated by biological systems, provide an example system in which spatial variation in elemental sources controls the isotopic composition of plants and animals. Work by Kennedy *et al.* (1998) on the well-characterized soil chronosequences of the Hawaiian islands nicely illustrates such a relationship. In this case, soils that are initially charged with Sr from the bedrock basalt exist in a dynamic balance where Sr lost through weathering and leaching from soils is replenished by Sr inputs from oceanic aerosols. Because the basalt and marine Sr sources have very different $^{87}Sr/^{86}Sr$ values, it is primarily land surface age, in turn determined by the age of the bedrock basalt flows, that is the primary determinant of Sr isotope ratios in soils and plants. The ages of basalt flows, both on the Big Island of Hawaii (Figure 4.1A) and throughout the island chain, are spatially patterned, creating a strong spatial patterning of Sr isotope ratios in Hawaiian island food webs.

In isotopic systems in which appreciable isotopic fractionation occurs in the environment and biological systems, spatial variation in the local environmental conditions under which isotope-fractionating processes occur can have a large impact on the isotopic composition of plants and animals. The photosynthetic assimilation of carbon by plants, for example, is a highly fractionating process that uses carbon from a relatively isotopically homogeneous environmental substrate, atmospheric CO_2. As such, there is relatively little potential for spatial isotopic variation in elemental sources to drive variation in plant $\delta^{13}C$ values (with exceptions such as near urban centers; Pataki *et al.* 2003), but great potential for variation in the *magnitude* of photosynthetic fractionation to produce spatial $\delta^{13}C$ variation. For C_3 ecosystems, photosynthetic discrimination is largely controlled by environmental parameters such as soil water availability, temperature, and atmospheric moisture content that influence the gas exchange physiology of leaves. As a result, strong relationships often exist between climate variables such as temperature or precipitation and plant or ecosystem $\delta^{13}C$, leading to spatial patterning of carbon isotope ratios that mimics variations in climate. This is nicely illustrated by measurements of the $\delta^{13}C$ of ecosystem respiration along a sampling transect in western Oregon (Figure 4.2; Bowling *et al.* 2002) that demonstrates a strong relationship between ecosystem carbon

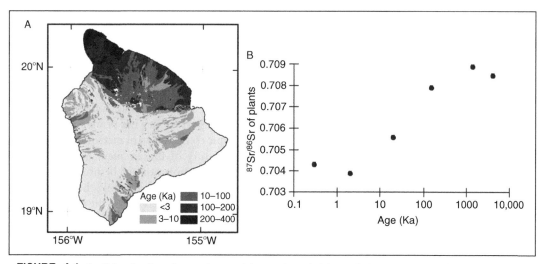

FIGURE 4.1 Spatial patterning of an environmental property underlying variation in ecosystem Sr isotope ratios on the Big Island of Hawaii. (A) The spatial distribution of approximate landscape surface ages (after Trusdell *et al.* 2006) on the Big Island of Hawaii. (B) A strong relationship exists between surface age and plant $^{87}Sr/^{86}Sr$, reflecting the decreased availability of basalt-derived Sr and increased accumulation of Sr from sea-salt aerosols with greater surface age (Kennedy *et al.* 1998).

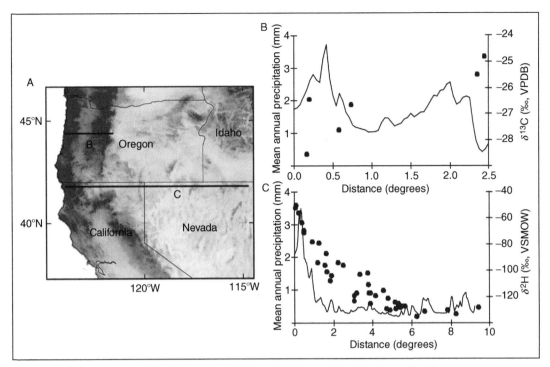

FIGURE 4.2 Landscape-level stable isotope variation related to spatial patterning of climatic gradients. (A) Strong gradients in mean annual precipitation amount (Daly *et al.* 1994) characterize two western US isotope sampling transects (bold lines). (B) Carbon isotope ratios of ecosystem respiration (dots, estimated from Keeling plots; Bowling *et al.* 2002) show a strong relationship to position along the Oregon transect, and are broadly, negatively correlated with precipitation amount (line). This reflects the strong relationship between local fractionation due to plant gas exchange processes and climatology along the spatial climate gradient. (C) Hydrogen isotope ratios of meteoric water samples (groundwater, springs, and streams; Ingraham and Taylor 1991) varying continuously along a transect in northern California and Nevada are positively correlated with changes in precipitation amount along the coast to interior gradient. The gradients in the water isotopic composition and precipitation amount can each be related to variation in the processes governing air mass water balance and transport along the dominant westerly atmospheric circulation trajectory for this region.

isotope ratios and precipitation amount along a strong, coast-to-continent gradient in precipitation amount.

The third, and most complex, type of relationship linking spatial isotopic variation to spatial variation in environmental factors is found in cases where transport history is a primary determinant of isotopic composition. We have seen, for example, the isotopic composition of meteoric precipitation is dependant, among other factors, on the condensation history of the air mass from which the precipitation forms. As a result, δ^2H and $\delta^{18}O$ values of precipitation can not be predicted *a priori* without consideration of the trajectory taken by the atmospheric moisture, reaching the site at which the precipitation condenses. Spatial organization clearly exists at many levels within the climate system, however, including atmospheric circulation patterns and the temperature gradients that lead to condensation from air masses, leading to spatial patterning in transport and condensation history of precipitating air masses. Waters sampled along a gradient in mean annual precipitation amount from the northern California coast to eastern Nevada (Figure 4.2; Ingraham and Taylor 1991), for example, illustrate a strong, spatially determined relationship between precipitation amount and isotopic composition, which reflects the progressive drying of air masses leaving the northern Pacific Ocean and traversing western North America.

B. Two Classes of Isotopic Patterns

Spatial patterning of isotopic variation in many systems closely mimics that of the underlying environmental determinants. This isotopic patterning can take on a range of forms depending on the spatial distribution of variation in environmental parameters, and its form has important implications for the types of migration research applications to which each isotope system is suited. In this context, we distinguish two endmember classes: spatially continuous and spatially discrete patterns. Depending on the environmental determinants of spatial variation in a given isotope system and the spatial scale in question, isoscapes can depict isotopic variation that is spatially continuous, discrete, or intermediate to these endmember classes.

Spatially continuous isoscapes are characterized by smooth variation in isotope ratios across space, giving continuous isotopic gradients along map transects. These isoscapes occur in cases where the underlying environmental determinants vary continuously in space and the process through which they impact landscape-level isotope ratios is a continuous function. Continuous isoscapes are common but not ubiquitous where climatological factors drive isotope ratio variation: for example, the continuous (though not unidirectional) change in precipitation amount that is related to ecosystem $\delta^{13}C$ values in western Oregon or the progressive rainout of moisture during westerly circulation, driving precipitation δ^2H variation across northern California and Nevada (Figure 4.2).

Spatially discrete isoscapes represent the contrasting pattern of spatial variation in which isotope ratios of environmental substrates are relatively invariant over some areas but change abruptly and discretely across the boundaries between invariant zones. Such patterns of "patchy" spatial isotopic variation can occur where environmental drivers vary discretely across space or where continuously varying environmental factors influence isotopic variability according to discrete functions. Isotopic variation of Sr in Hawaiian island ecosystems (Figure 4.1) represents an excellent example of the first scenario, where the aerial extent of individual lava flows define land surfaces of common age, presumably characterized by similar ecosystem Sr isotope ratios, and the discrete boundaries of these flows segment the isotopic landscape. The second scenario can be illustrated by the case of natural, climate-induced variation in the distribution of C_3 and C_4 plants. Although driven by continuously varying climate parameters, the impact of climate on C_4 distribution (and thus ecosystem $\delta^{13}C$ values) is described by a threshold function (Ehleringer *et al.* 1997, Collatz *et al.* 1998), and in many cases the distribution of these plant types can be considered to be discrete, particularly over large spatial scales (Still *et al.* 2003). The issue of scale is highly relevant to the concept of discrete isoscapes, as both transport processes and statistical variance in populations tend to smooth the boundaries of discrete isoscapes, particularly at small spatial scales.

The recognition and distinction of isoscape pattern classes is critical for terrestrial migration ecology applications in that they largely determine the types of ecological questions to which different isotopic systems are applicable. Continuous isoscapes offer important, recognized opportunities for research in migratory connectivity, particularly at large spatial scales. Because isotope ratio variation within continuous isoscapes is largest along directional, environmental gradients, these isoscapes are most amenable to addressing questions dealing with the position of individuals along geographic gradients. The continuous isotope ratio variation across these isoscapes can confound attempts to apply them to questions of explicit assignment of individuals to discrete locations, although statistical assignment techniques have been successfully applied to questions that can be posed as assignment to regions (Chapter 5; Wunder *et al.* 2005). Discrete isoscapes are well suited in some cases to address assignment questions where the questions and constraints can be well defined and overlap with isotopically defined spatial domains. However, isotope systems that are characterized by discrete spatial patterning cannot be applied to assignment at spatial scales below that of the patches comprising the isoscape, and depending on the environmental forcers involved potential exists for nonunique assignments within the domain of study.

Published applications of isotopes to terrestrial migration research have capitalized on both classes of patterns. With the advent of hydrogen isotope ratios as a tool for wildlife forensics (Chamberlain *et al.* 1997, Hobson and Wassenaar 1997), much emphasis has been focused on the application of continuous isoscapes of $\delta^2 H$ in water to problems of migratory connectivity. However, several examples focusing on reconstructing the migration of modern or ancient animals and humans continue to exploit discrete spatial variability of isotope ratios, in particular those of Sr (Chamberlain *et al.* 1997, Hoppe *et al.* 1999). The relative strengths and weaknesses of each class of isoscape should be considered in future application, and in particular, we suggest that important opportunities exist for coupling isotope systems with continuous and discrete isoscapes characterized by variation at different spatial scales.

IV. MAPPING ISOSCAPES

Isoscape maps are created through a multistep process that is often iterative. Information on isotope fractionating processes and patterns, represented as models, is combined with geospatial data to produce site-specific isotope ratio predictions over a spatiotemporal domain of interest. Major steps in the process include model selection, model calibration, data acquisition, calculation, optimization of model residuals, and estimation of error.

A. Model Selection and Calibration

For many isotope systems, multiple models are available, or could be envisioned, to describe the fractionating processes determining spatial isotopic variability within the system. These may range in complexity and precision from purely theoretical first-principals equations to highly derived empirically calibrated functions. In many cases, trade-offs exist between the accuracy of a model and its ability to be generalized or applied over large spatial scales. Appropriate model selection is critical to obtaining an accurate and well-constrained isoscape for the system and spatiotemporal domain under consideration, and these factors together with the availability of data and intended uses of the product should guide the choice of model.

The least-specific classes of models that can be applied to any isotopic system are geostatistical interpolation models. These models describe the variation in isotope ratios in terms of location alone. Estimates are calculated as a function of observed values at nearby locations, meaning that the only data required are observations of the isotopic ratio at points within the spatial domain of interest. A range of interpolation models of varying complexity are available, and some of the simpler examples such as triangulation, nearest neighbor, and inverse distance techniques are widely applied by non-specialists. The only statistically "correct" method for spatial interpolation is known as kriging that is the name given to a class of procedures in which the weights assigned to nearby data during the interpolation process are determined by a model of the covariance structure of the observational data (Isaaks and Srivastava 1990). Interpolation models offer a convenient way of visualizing and extending observations from a well-developed sampling network across space, but the degree of detail in the resulting isoscape is entirely limited by that represented in the measured or observational data set. As a result, the use of interpolation models in isolation is best restricted to cases where the observational documentation of spatial isotopic variation is extensive. However, as we will see in the following sections, interpolation models can be used to improve isoscape accuracy when used in combination with other types of models.

More specialized models for spatially varying isotopic systems can be constructed using derived parameters as proxies for the environmental factors underlying the isotopic variation. These models are particularly useful in cases where the complexity or data requirements of first-principles or process-based models are prohibitive, but where a simpler or more readily obtained suite of surrogate parameters can be substituted in their place. Derived models have been used to create global isoscapes of precipitation isotope ratios by simplifying the extremely complex suite of environmental factors underlying spatial isotopic variation in precipitation to a derived model of the form:

$$\delta = a(L^2) + b(L) + c(A) + d, \tag{4.4}$$

where δ is the isotopic composition of precipitation, L is latitude, A is altitude, and a, b, c, and d are empirically fitted parameters (Bowen and Wilkinson 2002). As this example shows, the parameters used in derived models may be only indirectly related to the physical processes driving isotopic variation, but may be useful surrogates because of correlation with the first-order parameters. In most cases, parameter values will need to be calibrated relative to observational data, but if the model is stable and the parameters are skillfully chosen, some derived models will be capable of extrapolation to combinations of parameter values not represented in the observational data set.

A third class of models includes first-principles and process-based parameterizations. Because of the complexity of the processes underlying isotopic variability, there are few cases where models can be fully developed based on first principles alone. Where such models could be developed, it is likely that they would necessarily oversimplify the systems of interest, severely limiting their applicability. For example, a first-principles model of the radiogenic production of ^{87}Sr could be applied using data on rock ages and Rb contents to produce a bedrock $^{87/86}$Sr map, but this isoscape would likely have limited relevance to $^{87/86}$Sr of most terrestrial ecosystems because of reasons discussed above. In contrast, process-based models strive to faithfully represent the physical and chemical processes underlying isotopic variation but adopt parameterizations that may group or simplify the details of these processes. For example, commonly used models of photosynthetic ^{13}C fractionation by C$_3$ plants explicitly represent the biophysics of CO$_2$ gas exchange by leaves but incorporate species-specific or biome-specific parameters to describe the biological regulation of gas exchange by the stomata (Ball *et al.* 1987, Lloyd and Farquhar 1994). These model parameters must again be calibrated against field data, but given the more fundamental nature of these variables, this work can often be done through laboratory studies or experiments and extended to the spatiotemporal domain.

B. Geospatial Data

A mind boggling amount of geospatial data currently available to support scientific research are found in the form of digital data archives, reanalysis and data synthesis projects, GIS database and decision support tools, and real-time Earth-observing satellite data. Many types of geospatial data are relevant to isoscapes modeling, including physiographic information (*e.g.*, position data such as latitude and longitude, elevation, land surface slope), climate data (*e.g.*, temperature, precipitation amount, atmospheric humidity), geological data (*e.g.*, bedrock geological maps, soil types), hydrological data (*e.g.*, stream routing, aquifer distribution and depth), biological data (*e.g.*, species or biome distributions, leaf area index, the normalized difference vegetation index), and socioeconomic and demographic data (*e.g.*, distribution of crop production, population density, land use).

Major data distributors include the World Data Center System, the National Aeronautics and Space Administration, the National Center for Atmospheric Research, Oak Ridge National Laboratory Distributed Active Archive Center, the National Oceanic & Atmospheric Administration, the US Geological Survey, the Climatic Research Unit, the European Centre for Medium-Range Weather Forecasts, and synthesis projects such as The International Satellite Land-Surface Climatology Project. Depending on the data type and provider, differing levels of spatial and temporal coverage may be available: many data sets are available for individual states or countries, and the temporal sampling interval of satellite-gathered data may be widely different from ground-based observational products. These issues must be taken into account and reconciled, for example, through creating mosaics of data from multiple sources or averaging data over a common time window, before the data can be used in isoscapes models.

Because the end goal of isocapes modeling is to produce continuous surfaces of isotopic variation in space, the data input to these models must be amenable to representation in raster format. A raster is a representation of data in two or more dimensions (*e.g.*, latitude and longitude for many geospatial rasters) as a grid of regularly spaced cells, each containing a single value. Rasters are a limited data format, in that each grid can only contain values for a single parameter, but they are useful for isoscapes modeling in that they are continuous (*i.e.*, they provide values or "no data" indicators for the entire spatial extent of the raster) and many rasters can be formatted uniformly to allow them to be combined and analyzed together. Many geospatial data are distributed in raster format, and others can be easily converted (*e.g.*, a polygon that represents the aerial extent of a biome can be converted to a raster grid of 1s and 0s, representing the presence or absence of the biome across a spatial domain). The geometry of rasters is described in terms of their extent (what are the boundaries of the data?), resolution (what is the size of an individual grid cell?), and projection (what coordinate system is used to describe the spatial relationship between cells?). To allow accurate calculation of isoscapes, raster data must be processed so that each of these properties is uniform across all data sets.

Finally, many isoscapes models require spatially distributed observations of isotope ratios as input or for calibration purposes. A few outstanding spatial observation networks have produced valuable spatially resolved isotopic data sets, including several IAEA programs such as the Global Network of Isotopes in Precipitation and Rivers, the US National Oceanic and Atmospheric Administration's Cooperative Air Sampling Network, and the US Geological Survey's North American Stream Quality Accounting Network. In a few cases, significant, spatially extensive (although temporally limited) data collections have resulted from the investigator-driven research (*e.g.*, Longinelli and Selmo 2003, Bowen *et al.* 2007), but in most cases, a high degree of coordination among multiple investigators is needed to obtain, analyze, and manage such data sets. A general lack of networked observation programs for isotopes places unfortunate limits on the availability of data that could greatly advance the quality and scope of isoscapes for migration research.

C. Model Calculations

Following model selection, calibration, and data assembly, a provisional isoscape is created by executing the model calculations on a spatial grid using spatial data input layers. In most cases, this procedure is simply that of iteratively executing a set of calculations that solve the model equations at each grid cell in the spatial domain. Model calculations thus require a computational routine that iterates through the cells, reads data for that cell from the required data rasters, executes the model calculations, and outputs the result to a new output raster. These functions can be accomplished through hand-coded routines consisting of iterative loops, file input/output statements, and mathematical operators written in any programming language that supports these functions (*e.g.*, BASIC, FORTRAN, C, C++, Java). They can also be implemented in most commercial GIS software packages (ArcGIS, GMT, GRASS, GrADS, SURFER) that provide much of the basic functionality needed to execute these routines in the form of prebuilt tools that accomplish many of the lower-level functions (*e.g.*, reading a raster data file or iterating through the spatial grid) transparently. In many cases, this can increase the efficiency of the calculation and routine-building process, and because the GIS tools are typically available through graphical user interfaces, it allows isoscape calculations to be conducted by researchers having little or no computer programming experience.

In addition to the actual process of model execution, important decisions must be made at this step about the extent and resolution of the isoscape that will be generated: how large an area will be modeled and at how fine a spatial division? The question of extent may be one that is answered largely by practical constraints, for example, the aerial coverage required to encompass the likely range of a migrant population or the extent of a requisite geographic data set. In making decisions about the spatial extent of modeling, however, it is important to consider the extent of any calibration data that were used and critically assess whether they are sufficient to support application of the isoscapes model at the desired extent or whether unwarranted extrapolation will be involved.

Decisions about the spatial resolution at which model calculations are applied are in many cases more subjective but are equally important. A number of factors must be considered in the selection of an appropriate spatial resolution for a particular isoscape and ecological application. Biological factors, for example, the size of an individual's range on the breeding or wintering grounds, will be relevant to determining the maximum resolution that is appropriate for a particular application. In a more general sense, the number, density, and geographic specificity of calibration data will have a strong influence on the degree of spatial specificity that is appropriate for the modeling work. In this regard, it is important for both modelers and isoscapes users to recognize that higher spatial resolution is not necessarily better. In many cases, attempts to predict isotope distributions at very high spatial resolutions may actually compromise the overall quality of the resulting data products because the resolution at which the calibration and calculation work is conducted may exceed that of the physical spatial processes determining isotopic variability and introduce artifactual or overspecified dependence of the predicted isotope ratios on model variables. There is some evidence that this may be the case for some published high-resolution isoscapes of precipitation isotopic composition (Chapter 5).

D. Optimization of Residuals

Models, as simplified descriptions of the processes they represent, inevitably fail to make perfect predictions. This is true for isoscapes models that are often limited in their complexity by the availability of spatial input data that would allow their application over the spatial scales of interest. Mismatch between the model predictions and observational data, or model residuals, can often be partly attributed to uncertainties in the data or the model parameters, but may carry important information about the inadequacies of the model. Where spatial data sets documenting the isotopic values of interest at points within the modeling domain are available, isoscape modeling offers two powerful ways to take advantage of residuals in order to optimize the end data product.

First, examination of model residuals can often produce insight into the model through comparison with ancillary data sets, leading to iterative improvements in the modeling. The residuals can be considered test cases, and mismatch with the observational data may highlight missing processes that could be incorporated to improve future modeling. GIS software facilitates the comparison of residual values with the wide range of available spatial data by allowing users to intersect and extract data at specified locations from multiple data sets: for example, residual values from a network of isotope-observing sites could be referenced against data showing land use categories to produce a spreadsheet. comparing these data across all sites.

Second, in many cases, residual values are nonrandomly distributed in space. This spatial autocorrelation presumably reflects a location-dependant process that was incompletely represented in the model, and analysis of residual spatial patterning can itself offer insight into missing model processes. Spatial autocorrelation offers another avenue to improving isoscape accuracy, however, through the use of a geostatistical interpolation model. The geostatistical model can be applied to interpolate a grid of predicted residual values that can be added to the isoscape model predictions to "correct" them against the observational data. An early analysis of precipitation isotope ratios by Bowen and Wilkinson (2002) illustrates the use of spatially autocorrelated residuals (Figure 4.3). In this case, spatially patterned residuals suggested that aspects of the climatology and atmospheric circulation not represented by the derived isoscape model used in the study had a significant impact on precipitation isotope ratios.

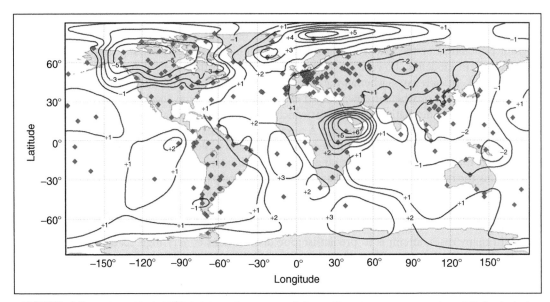

FIGURE 4.3 Model residuals ($\delta^{18}O$, ‰) for an isoscape predictions of long-term, mean annual precipitation isotopic composition (reprinted from Bowen and Wilkinson 2002). Residual values were calculated as observed values—model predictions at precipitation monitoring sites. Low residuals values over the northern hemisphere continental interiors can be attributed to the extensive rainout of ^{18}O as air masses traverse the continents. High residuals over the mid- and high-latitude oceans can be attributed to extensive evaporation from the warm surface waters of the gyres in these regions. The large magnitude positive residuals in eastern Africa reflect a combination of factors including proximity to Indian Ocean water sources, reevaporation of falling rain, and relatively low decrease of isotopic values with elevation in this region.

This information could be used to develop an improved model parameterization that incorporated these environmental factors, or, as was done in the cited study, to develop a residual correction using geostatistical interpolation.

E. Uncertainty of Isoscape Predictions

As we have shown, isoscapes are commonly the product of multilevel modeling efforts involving derived spatial data products, imperfect model parameterizations, and geostatistical analysis. Developing quantitative measures of the uncertainty of isoscapes predictions represents a current and significant challenge to the robust application of these products to migration research. Prediction error defined as the difference between an isoscape prediction at a given site and the true value of the modeled variable is itself spatially variable, and so the most useful data products documenting uncertainty are maps of standard errors, confidence intervals, or related statistics across the modeling domain. Relatively few such maps have yet been presented in the literature (*e.g.*, see Bowen and Revenaugh 2003). It is important to note that prediction error as discussed here represents only one component of error affecting ecological interpretations developed using isoscapes. Selection of appropriate isoscapes for application to a particular question (*e.g.*, annual average vs seasonal precipitation; Bowen *et al.* 2005) and development of accurate models for the relationship between tissue isotopic composition and that of the environmental substrates modeled by isoscapes are important sources of uncertainty that are beyond the scope of this chapter.

Prediction error can be partitioned into error from two sources: data error and model error. Sources of error in the data products used to calibrate, drive, or optimize spatial stable isotope models can be related to the handing and analysis of individual samples (*e.g.*, evaporation of water from improperly stored samples prior to H and O isotope ratio analysis), lack of homogeneity within large data sets (*e.g.*, using data having uneven temporal coverage to estimate long-term average isotopic values), or the generation of derived data products (*e.g.*, errors in climate model reanalyses or interpolated climate rasters). Model errors relate to the parameterization of natural processes used for modeling and the inaccuracies in these parameterizations. For example, maps of precipitation isotope ratios were produced by Bowen and Wilkinson (2002) using a fixed coefficient for variation with altitude [parameter c in Eq. (4.4)] at all sites, a simplifying and inaccurate assumption that biased predictions in parts of East Africa. The distinction between data and model error is often blurred in isoscapes modeling because many of the data products used are themselves model-derived.

Several methods are available for quantification of isoscapes prediction uncertainty, all of which provide accurate assessment of data error but may incompletely represent model error. Metrics of error can be derived from the covariance matrix in kriging or through error propagation (assuming estimates of parameter and variable uncertainties are available) in process-based and derived-parameter models. Resampling statistics (*e.g.*, cross-validation, jackknife, and bootstrap methods; Wu 1986) can also be applied to generate estimates of uncertainty in any of these cases, and while these methods are computationally intensive they have the significant advantage of being insensitive to assumptions about probability distribution functions and covariance functions of model parameters (*i.e.*, they are nonparametric methods). Each of these methods will quantify error associated with noisy data by calculating its impact of the precision with which model parameters are known. Model error, however, will only be adequately represented where data are available to document its impact on the precision of the model. In the case of East African precipitation isotope ratios given above, for example, the subsequent error analysis of Bowen and Revenaugh (2003) was able to identify high uncertainty for estimates in this region based on the data from a monitoring station in this region (Addis Abba). In the absence of local data, however, the inability of the fixed altitude effect parameterization to correctly predict isotope ratios in this region would not have been recognized. In such cases, the only clear routes to improving isoscape models and the estimation of error are through the parallel paths of expanded spatial data collections and improved model parameterization.

V. ISOSCAPES FOR TERRESTRIAL MIGRATION RESEARCH

In this final section, we review many of the currently available isoscapes data products as they relate to the study of terrestrial animal migration.

A. H and O Isoscapes

Isoscapes of H isotopes in water, particularly in precipitation, have been widely applied in migration research based on the premise that the dominant source of hydrogen in body tissues is environmental water, either consumed directly or routed through diet (*e.g.*, Chamberlain *et al.* 1997, Hobson and Wassenaar 1997, Norris *et al.* 2004, Bearhop *et al.* 2005). Analogous application of O isoscapes has been investigated in at least one case (Hobson *et al.* 2004), but so far has shown less promise. The most commonly referenced products are maps of long-term average growing season precipitation isotope ratios (Figure 4.4; *e.g.*, Meehan *et al.* 2004, Bowen *et al.* 2005) produced by derived-parameter modeling of monthly data with geostatistical residual correction. These products are freely available on the web, and although uncertainty estimates specific to the growing season products have not been produced, confidence intervals generated for related annual average precipitation isoscapes (Bowen and Revenaugh 2003) provide a general indication of the potential error in the growing season maps.

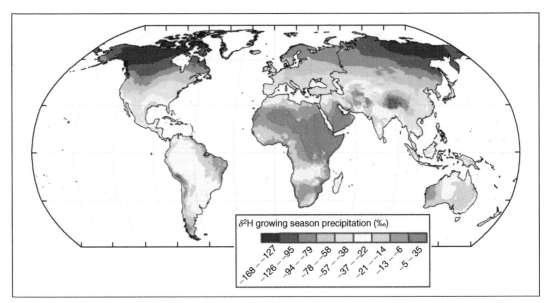

FIGURE 4.4 Global isoscape of long-term average growing season precipitation δ^2H values (after Bowen *et al.*, 2005). This map represents the precipitation amount-weighted average of monthly isoscapes produced using a derived parameter model [modified version of Eq. (4.4)] and interpolated residual correction. Climate data (temperature and precipitation) was interpolated from the Global Historical Climatological Network (Peterson and Vose 1997). **(See Color Plate.)**

The nature and extent of isotopic variability across these isoscapes has been reviewed by Bowen *et al.* (2005). In general, they provide the greatest power to constrain the location of migration endpoints and pathways within the northern hemisphere continental interiors, and in some cases may also be useful for differentiating habitats along altitudinal gradients.

One of the greatest limitations of currently available precipitation-based isoscapes is their limited temporal specificity. In order to achieve high levels of accuracy in precipitation isoscape modeling, it has thus far been necessary to work with long-term average data accumulated over many decades of monitoring and measurement. Although the use of large, long-term average data sets greatly increases the accuracy of long-term average precipitation isoscapes (Bowen and Revenaugh 2003), it is not clear how closely these predictions reflect the environmental isotopic signal taken up by migrant individuals that inhabit a location for a period of weeks or months during a particular dry or wet year. Assessment of the variability of precipitation isotope ratios among multiple years at given sites is not a simple task, given the inconsistent and often severely limited temporal coverage of currently available data, but is needed in order to better understand the impact of year-to-year variability on the precision of interpretations made using precipitation H and O isoscapes. A similar challenge exists with respect to the seasonal specificity of these isoscapes. The rationale for restricting the data used to the "growing season" (*e.g.*, defined as all months with mean temperature $>0\,°C$) has been that water assimilated by plants and entering the food chain will be primarily derived from growing-season precipitation, but other seasonal or annual isoscapes may be more relevant in some cases (*e.g.*, Bowen *et al.* 2005). In this case, analysis and maps of the intra-annual variability in precipitation isotopic composition are available to document the potential effects on applications to migration research (Bowen in review).

For some migratory birds and animals, H or O isoscapes of precipitation may be a poor choice for isotope tracking work, and in these cases, a new generation of isoscapes representing other water sources may improve and expand the use of water isotopes in migration research. Isoscapes of surface (river) water (Fekete *et al.* 2006) and human tap water (Bowen *et al.* 2007) have now been produced for the contiguous United States, and may be relevant to tracing some migrants, for example, with aquatic habits or that might be known to feed primarily in irrigated agricultural habitats. These products have been developed using precipitation isoscapes to represent the isotope flux to the land surface and modeling the modification of this signal using hydrological (surface water) or geostatistical (tap water) models. The first-order patterns of these isoscapes are similar to those of precipitation, but the surface and tap water isoscapes demonstrate significant differences relative to precipitation products particularly in mountainous regions and along the course of large, mountain-fed rivers. Estimates of the accuracy of these products are preliminary, but comparisons of the river water product of Fekete *et al.* (2006) to river water monitoring data suggests that the significant errors that exist could be corrected using residual interpolation to produce a river water H isoscape of relatively high accuracy.

B. Vegetation C Isoscapes

Carbon isotope ratios have been primarily applied to studies of animal diet, largely because of the distinct isotopic signal associated with the relative proportion of C_3 versus C_4 plants in the diet (MacFadden and Cerling 1994, Cerling *et al.* 1997a,b, Sponheimer and Lee-Thorp 1999, Peters and Vogel 2005). They have not been as widely used to study migration (aquatic organisms being an important exception to this), perhaps due to disagreement in the literature over the utility of $\delta^{13}C$ in yielding geographic information (Chamberlain *et al.* 2000, Wassenaar and Hobson 2001, Hobson *et al.* 2003). Vegetation $\delta^{13}C$ isoscapes have significant potential to provide important insights by allowing one to compare, for example, spatially distributed data with continuous grid predictions or by providing an interpretation platform for bird feathers produced in different but unknown locations. Significant spatial information may therefore be found from model predictions that incorporate not only the distribution of C_3 and C_4 plants but also the comprehensive isoscapes that include these distributions as well as the climatic responses. By combining models of C_3 versus C_4 plant distributions, biophysical models of plant carbon isotope fractionation, and biome distributions derived from satellites to allow biome-specific plant physiology to be incorporated, global plant $\delta^{13}C$ isoscapes have been produced (see Figure 4.5; Lloyd and Farquhar 1994, Scholze *et al.* 2003, Suits *et al.* 2005). These plant $\delta^{13}C$ isoscapes provide a useful framework for understanding observed spatial information in bird tissue $\delta^{13}C$ (Pain *et al.* 2004), assuming one understands the relationships between birds and their food source. It is of course necessary also to have some confidence in the relationship between the isoscape prediction and the actual food source (*e.g.*, if the model predicts leaf $\delta^{13}C$ and the bird eats primarily seeds, what is the relationship between seed and leaf $\delta^{13}C$?). If these variables can be understood, plant $\delta^{13}C$ isoscapes offer the potential for sophisticated interpretations. Greater exploration of these interfaces is clearly warranted.

C. Vegetation N Isoscapes

Although mapping spatial variation in plant $\delta^{15}N$ has received less attention than has plant $\delta^{13}C$, there is at least one published set of global plant and soil $\delta^{15}N$ maps (Amundson *et al.* 2003). This modeling effort was based on prior arguments that plant $\delta^{15}N$ is related to the residence time of N in an ecosystem or N cycle "openness," as well as empirical observations consistent with this expectation that plant $\delta^{15}N$ is negatively correlated with precipitation (*e.g.*, Austin and Vitousek 1998, Handley *et al.* 1999). Temperature is also positively correlated with plant $\delta^{15}N$ values and this relationship is part of the model by Amundson *et al.* (2003) (see Figure 4.6). It has also been observed that animal $\delta^{15}N$ is negatively correlated with precipitation, a pattern that could be caused by both changes in dietary $\delta^{15}N$ or animal metabolism. Recent work suggests that the pattern is the result of variation in dietary $\delta^{15}N$ and not variation in animal metabolism indicating that animals (in this case, kangaroos)

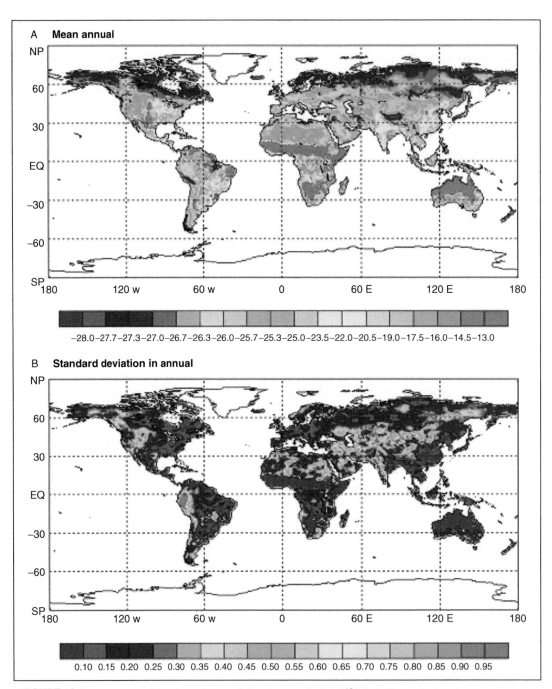

FIGURE 4.5 Global mean (A) and standard deviation (B) annual plant $\delta^{13}C$ (‰). These plant carbon isoscapes are hybrid products derived from global distribution maps of C3/C4 vegetation (derived from satellite products and physiological modeling) and modeled physiological responses of C3 plants to atmospheric conditions for the years 1983–1993 (continuous fields from ECMWF) and constrained by the Normalized Difference Vegetation Index for those years. Figure reproduced with permission from Suits *et al.* (2005). **(See Color Plate.)**

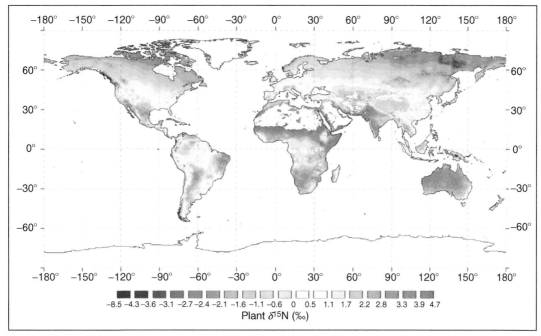

FIGURE 4.6 Global plant $\delta^{15}N$. This plant nitrogen isoscape was produced by executing a regression model in GIS previously fit to observed plant $\delta^{15}N$ and mean annual precipitation and temperature (Amundson *et al.* 2003). The model was driven with observed climate parameters (continuous fields of Mean Annual Precipitation and Mean Annual Temperature for the climate normal period 1961–1990 from the Climate Research Unit; New *et al.* 2002). The model output was further masked using continuous vegetation fields (DeFries *et al.* 1999) eliminating areas with greater than 80% nonvegetated ground. **(See Color Plate.)**

faithfully record dietary $\delta^{15}N$ and that dietary $\delta^{15}N$ is itself linked to climate, likely through its effect on N cycle openness (Murphy and Bowman 2006). Support for retention of this geographic signal has also been found for warblers (Chamberlain *et al.* 2000). As with carbon, however, explicitly incorporating plant $\delta^{15}N$ isoscapes into the study of animal migration remains largely unexplored and should provide important insights as these efforts increase.

D. Vegetation H and O Isoscapes

Because hydrogen and oxygen are found in plant organic compounds and water, both plant water and organic isoscapes have been produced for application to understanding biosphere–atmosphere interactions, and others such as to commerce or forensics (see Figure 4.7; Farquhar *et al.* 1993, Ciais *et al.* 1997, Cuntz *et al.* 2003, West *et al.* 2007). These plant isoscapes have, however, not yet been widely applied to improving understanding of migration, relying instead on strong relationships between animal isotope ratios (primarily H), and drinking water isotope ratios. It seems likely that a greater understanding of all significant sources of H and O inputs to animal metabolism would yield better predictive ability, making plant H and O isoscapes potentially useful for understanding animal movement. This is especially true for animals that obtain a significant amount of their body water from plant water. Plant δ^2H and $\delta^{18}O$ isoscapes are generally derived using a combination of approaches. Plant processes that discriminate against ^2H or ^{18}O are modeled explicitly using biophysical models of the fractionation. These models are themselves driven by parameters such as plant source water isotopic composition and climate that are either simulated within general circulation models or

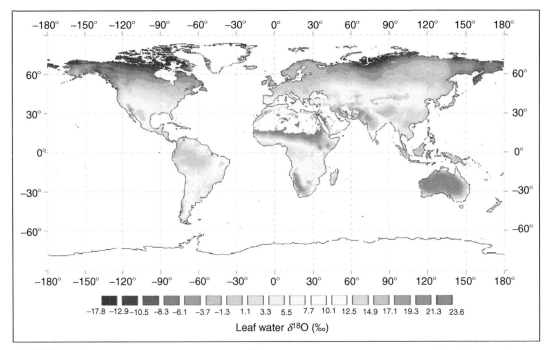

FIGURE 4.7 Global annual average leaf water $\delta^{18}O$. This plant oxygen isoscape was produced by executing a physiological model of leaf water enrichment (modified from Flanagan *et al.* 1991) using gridded annual average precipitation $\delta^{18}O$ (see Figure 4.3) and monthly climate parameters (continuous fields of temperature and relative humidity for the climate normal period 1961–1990 from the Climate Research Unit; New *et al.* 2002) to drive the model (West *et al.* in review). Monthly grids were averaged and the model output was further masked (as in Figure 4.5) using continuous vegetation fields (DeFries *et al.* 1999) eliminating areas with greater than 80% nonvegetated ground. **(See Color Plate.)**

derived from surface water or precipitation isosocapes. In addition, the resulting plant isoscapes may be then weighted using various approaches, including maps of plant biome distributions or productivity, or simulations of the same (West *et al.* in review). Depending on the degree of understanding or available data, the plant δ^2H and $\delta^{18}O$ isoscapes may be general, such as a global average leaf water isoscape, or quite specific, such as a series of isoscapes depicting the changing spatial variation of leaf water for a single growing season. The degree of detail may be dictated by the availability of data or model understanding, or it may be dictated by the specificity or generality of the question being asked. As with all isoscapes discussed, the approach is flexible and not *a priori* linked to any particular temporal or spatial scale.

E. Ecosystem Sr Isoscapes

The only published example of a large-scale Sr isoscape (Beard and Johnson 2000) represents the bedrock $^{87}Sr/^{86}Sr$ of the contiguous United States as a function of rock age (Figure 4.8). As discussed previously, local bedrock Sr isotope ratios in many cases will be only loosely related to ecosystem $^{87}Sr/^{86}Sr$ relevant to migration research applications (Naiman *et al.* 2000), but for large-scale, low-resolution tracking efforts, the patterns predicted by the Sr isoscape likely provide a reasonable template for first-pass interpretations. In particular, the areas of high $^{87}Sr/^{86}Sr$ in areas of very old (Precambrian) bedrock in northern Minnesota and some Rocky Mountain states likely represent areas

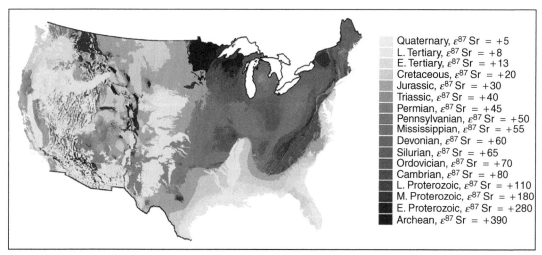

Quaternary, ε^{87} Sr $= +5$
L. Tertiary, ε^{87} Sr $= +8$
E. Tertiary, ε^{87} Sr $= +13$
Cretaceous, ε^{87} Sr $= +20$
Jurassic, ε^{87} Sr $= +30$
Triassic, ε^{87} Sr $= +40$
Permian, ε^{87} Sr $= +45$
Pennsylvanian, ε^{87} Sr $= +50$
Mississippian, ε^{87} Sr $= +55$
Devonian, ε^{87} Sr $= +60$
Silurian, ε^{87} Sr $= +65$
Ordovician, ε^{87} Sr $= +70$
Cambrian, ε^{87} Sr $= +80$
L. Proterozoic, ε^{87} Sr $= +110$
M. Proterozoic, ε^{87} Sr $= +180$
E. Proterozoic, ε^{87} Sr $= +280$
Archean, ε^{87} Sr $= +390$

FIGURE 4.8 Predicted ^{87}Sr/^{86}Sr (as ε^{87}Sr $= [(^{87}$Sr/^{86}Sr$)_{predicted}/(^{87}$Sr/^{86}Sr$)_{bulk\ earth} - 1] \times 10{,}000$) for bedrock of the contiguous United States (Beard and Johnson 2000). Values were modeled by assuming a fixed initial ^{87}Sr/^{86}Sr and ^{87}Rb and Sr content and calculating ^{87}Sr production following rock formation using the known decay rate of ^{87}Rb and rock ages represented on the US Geological Survey digital geological map. Reprinted, with permission, from the *Journal of Forensic Sciences*, Volume 45, Issue 5, copyright ASTM International, 100 Barr Harbor Drive, West Conshohocken, PA 19428.

where Sr isotope ratios offer power to constrain the location and habitat of migrants. Relatively low ^{87}Sr/^{86}Sr mapped along the Gulf coastal plain is also consistent with the smaller-scale study of Hoppe *et al.* (1999), which, however, demonstrated additional fine structure in the spatial isotope patterning by mapping plant and rodent tooth Sr isotope ratios at sites across northern Florida and southern Georgia. The structure can largely be attributed to differences in rock type (carbonate vs siliciclastic vs igneous/metamorphic), a factor not included in the Beard and Johnson model but which could be incorporated in improved Sr isoscapes models. In regions where atmospheric deposition has been identified as an important mechanism of Sr addition to the landscape, Sr isoscapes might also be improved by basing predictions on measurements of dust ^{87}Sr/^{86}Sr rather than predicted local bedrock values (Naiman *et al.* 2000).

VI. SUMMARY AND LOOK FORWARD

The current generation of isoscapes represents an attempt to reproduce the natural fingerprint of stable isotopes on the landscape by synthesizing geospatial data and models representing the processes underlying spatial isotopic variation. The organization of these processes in the natural and anthropogenically modified environment produces patterns of spatial variation that differ among isotope systems and substrates. Each of the isotope systems discussed here (H, C, N, O, Sr) as well as the sulfur isotope system is characterized by some level of known spatial variability and thus presents opportunities for application to migration research. The greatest potential for precise tracking of migratory connectivity using isotopes may exist where combinations of discrete and continuous variation occur at nested spatial scales. Isoscapes are produced using models and data of varying complexity and specificity. All of these models and data are imperfect and accurate estimates of prediction uncertainty and continued improvements in data and models are needed to advance the scope and quality of isoscapes data products available to migration researchers. Moreover, data products having greater specificity with respect to substrate (*e.g.*, lake vs river vs precipitation water isoscapes; leaf water isoscapes for different plant types) and spatiotemporal domain are needed to reduce the reliance of migration researchers on generalized products that may be of limited (and poorly known) relevance to their system of study.

The development and testing of isoscapes models is an exciting and new area of research, and improvements in these models can be expected as research in this field advances in the coming years. In contrast, new efforts in at least two key areas will be needed to improve the quality and availability of isoscapes data products for migration research. First, the quality of all isoscape predictions is dependent at some level on the availability of spatially distributed isotopic data. Whether used only for model calibration or as input to geostatistical models, the quality and coverage (*e.g.*, spatiotemporal, physiographic, and climatological) of this data is critical to assuring the accuracy of isoscapes data products. Continued and renewed effort to collect regular, spatiotemporally distributed isotopic measurements of relevant environmental substrates is a critical area of need for the advancement of isoscapes modeling. These efforts require coordinated sampling and analysis of large numbers of samples from widely distributed locations, and are thus almost always beyond the ability of single investigators or research groups and require the coordination of national or international networks. Relatively few such programs exist, and these have generated the majority of data used for water isoscapes modeling (*e.g.*, the GNIP), and produced large data sets of atmospheric gas isotope ratios (*e.g.*, the National Oceanographic at Atmospheric Administration's Earth System Research Laboratory Global Monitoring Division). Promise for the future advancement of network-based data gathering exists in new IAEA water sampling programs for major rivers (the Global Network for Isotopes in Rivers) and atmospheric water vapor (Moisture Isotopes in the Biosphere and Atmosphere) and in efforts to develop coordinated nation-wide scientific measurement platforms in the United States (*e.g.*, the WATERS network and National Ecological Observation Network), which could provide the infrastructure and coordination for future spatiotemporal isotope data gathering.

A second area where emphasis and transformation is required is in the current model of isoscapes generation and distribution. As isoscape data products of greater specificity are sought for ecological applications, the current model of research-driven generation and distribution of static data products will no longer suffice: it will not be possible for a researcher to produce and distribute the nearly infinite catalogue of data products required for the diverse and growing array of applications. In moving forward, emphasis must shift from the sharing of data products to the sharing of models. Here, cyberinfrastructure provides a wealth of untapped opportunities. High-speed data networks, grid computing (through which computationally intensive tasks can be executed using processor resources at distributed locations), the increased availability of data on the web (including real-time satellite data), and the development of software tools that can tap all of these resources "behind the scenes" ("middleware") provide the necessary infrastructure for web portals where isoscape developers can share their models and algorithms with the broader community. If these resources are properly integrated, it will be possible to create a web resource where, for example, a migration ecologist having no specialized knowledge of plant physiology, remote sensing, or geostatistics could produce and explore isoscapes of leaf water δ^2H values and their uncertainty, generated using plant physiology models, satellite data, and geostatistical analysis, for the specific spatiotemporal domains and species of relevance to his/her research project. More than any other advance, the development of such tools may promote the widespread exploration of isoscapes and invigorate the testing and application of isoscapes in migration research.

VII. ACKNOWLEDGMENTS

We thank Neil Suits and Brian Beard for providing isoscapes from their work reproduced here.

VIII. REFERENCES

Amundson, R., A. T. Austin, E. A. G. Schuur, K. Yoo, V. Matzek, C. Kendall, A. Uebersax, D. Brenner, and W. T. Baisden. 2003. Global patterns of the isotopic composition of soil and plant nitrogen. *Global Biogeochemical Cycles* **17**(1):1031.

Austin, A. T., and O. Sala. 1999. Foliar $\delta^{15}N$ is negatively correlated with rainfall along the IGBP transect in Australia. *Australian Journal of Plant Physiology* **26**:293–295.

Austin, A. T., and P. M. Vitousek. 1998. Nutrient dynamics on a precipitation gradient in Hawaii. *Oecologia* **113**(4):519–529.

Ball, J. T., I. E. Woodrow, and J. A. Berry. 1987. A model for predicting stomatal conductance and its contribution to the control of photosynthesis under different environmental conditions. Pages 221–224 *in* J. Biggins (Ed.) *Progress in Photosynthesis Research.* Martinus Nijhoff, Dordrecht.

Beard, B. L., and C. M. Johnson. 2000. Strontium isotope composition of skeletal material can determine the birth place and geographic mobility of humans and animals. *Journal of Forensic Science* **45**(5):1049–1061.

Bearhop, S., W. Fiedler, R. W. Furness, S. C. Votier, S. Waldron, J. Newton, G. J. Bowen, P. Berthold, and K. Farnsworth. 2005. Assortative mating as a mechanism for rapid evolution of a migratory divide. *Science* **310**:502–504.

Birks, S. J., J. J. Gibson, L. Gourcy, P. K. Aggarwal, and T. W. D. Edwards. 2002. Maps and animations offer new opportunities for studying the global water cycle. Eos, Transactions, American Geophysical Union, Electronic Supplement 83, http://www.agu.org/eos_elec/020082e.html.

Bowen, G. J. Spatial analysis of the intra-annual variation of precipitation isotope ratios and its climatological corollaries. JGR, in press.

Bowen, G. J., and J. Revenaugh. 2003. Interpolating the isotopic composition of modern meteoric precipitation. *Water Resources Research* **39**:1299.

Bowen, G. J., and B. Wilkinson. 2002. Spatial distribution of $\delta^{18}O$ in meteoric precipitation. *Geology* **30**(4):315–318.

Bowen, G. J., L. I. Wassenaar, and K. A. Hobson. 2005. Global application of stable hydrogen and oxygen isotopes to wildlife forensics. *Oecologia* **143**(3):337–348.

Bowen, G. J., J. R. Ehleringer, L. A. Chesson, E. Stange, and T. E. Cerling. 2007. Stable isotope ratios of tap water in the contiguous USA. *Water Resources Research* **43**:W03419.

Bowling, D. R., N. G. McDowell, B. J. Bond, B. E. Law, and J. R. Ehleringer. 2002. ^{13}C content of ecosystem respiration is linked to precipitation and vapor pressure deficit. *Oecologia* **131**:113–124.

Capo, R. C., B. W. Stewart, and O. A. Chadwick. 1998. Strontium isotopes as tracers of ecosystem processes; theory and methods. *Geoderma* **82**(1–3):197–225.

Cerling, T. E., J. M. Harris, S. H. Ambrose, M. G. Leakey, and N. Solounias. 1997a. Dietary and environmental reconstruction with stable isotope analyses of herbivore tooth enamel from the Miocene locality of Fort Ternan, Kenya. *Journal of Human Evolution* **33**:635–650.

Cerling, T. E., J. M. Harris, B. J. MacFadden, M. G. Leakey, J. Quade, V. Eisenmann, and J. R. Ehleringer. 1997b. Global vegetation change through the Miocene/Pliocene boundary. *Nature* **389**(6647): 153–158.

Cerling, T. E., J. M. Harris, and B. H. Passey. 2003. Diets of East African Bovidae based on stable isotope analysis. *Journal of Mammalogy* **84**(2):456–470.

Chamberlain, C. P., J. D. Blum, R. T. Holmes, X. H. Feng, T. W. Sherry, and G. R. Graves. 1997. The use of isotope tracers for identifying populations of migratory birds. *Oecologia* **109**(1):132–141.

Chamberlain, C. P., S. Bensch, X. Feng, S. Akesson, and T. Andersson. 2000. Stable isotopes examined across a migratory divide in Scandinavian willow warblers (Phylloscopus trochilus trochilus and Phylloscopus trochilus acredula) reflect their African winter quarters. *Proceedings of the Royal Society of London Series B-Biological Sciences* **267**(1438):43–48.

Ciais, P., A. S. Denning, P. P. Tans, J. A. Berry, D. A. Randall, J. J. G. Collatz, P. J. Sellers, J. W. C. White, M. Trolier, H. J. Meijer, R. J. Francey, P. Monfray, *et al.* 1997. A three dimensional synthesis study of $\delta^{18}O$ in atmospheric CO_2 Part 1: Surface fluxes. *Journal of Geophysical Research* **102**(D5): 5857–5872.

Collatz, G. J., J. A. Berry, and J. S. Clark. 1998. Effects of climate and atmospheric CO_2 partial pressure on the global distribution of C-4 grasses: Present, past, and future. *Oecologia* **114**(4):441–454.

Craig, H. 1953. The geochemistry of the stable carbon isotopes. *Geochimica et Cosmochimica Acta* **3**:53–92.

Craig, H. 1961. Isotopic variations in meteoric waters. *Science* **133**:1702–1703.

Craig, H., and L. I. Gordon. 1965. Deuterium and oxygen-18 variations in the ocean and the marine atmosphere. *in* E. Tongiorgi (Ed.) *Proceedings of a Conference on Stable Isotopes in Oceanographic Studies and Paleotemperatures.* Spoleto, Italy.

Cuntz, M., P. Ciais, G. Hoffmann, and W. Knorr. 2003. A comprehensive global three-dimensional model of $\delta^{18}O$ in atmospheric CO_2: 1. Validation of surface processes. *Journal of Geophysical Research* **108**(D17):4527.

Daly, C., R. P. Neilson, and D. L. Phillips. 1994. A statistical-topographic model for mapping climatological precipitation over mountainous terrain. *Journal of Applied Meteorology* **33**(2):140–158.

Dansgaard, W. 1954. The O^{18}-abundance in fresh water. *Geochimica et Cosmochimica Acta* **6**:241–260.

Dansgaard, W. 1964. Stable isotopes in precipitation. *Tellus* **16**:436–468.

DeFries, R. S., J. R. G. Townshend, and M. C. Hansen. 1999. Continuous fields of vegetation characteristics at the global scale at 1 km resolution. *Journal of Geophysical Research* **104**(16):11–16.

Dongmann, G., H. W. Nurnberg, H. Forstel, and K. Wagener. 1974. On the enrichment of H2O18 in the leaves of transpiring plants. *Radiation and Environmental Biophysics* **11**:41–52.

Ehleringer, J. R., and R. K. Monson. 1993. Evolutionary and ecological aspects of photosynthetic pathway variation. *Annual Review of Ecology and Systematics* **24**:411–439.

Ehleringer, J. R., P. W. Rundel, and K. A. Nagy. 1986. Stable isotopes in physiological ecology and food web research. *Trends in Ecology & Evolution* **1**:42–45.

Ehleringer, J. R., T. E. Cerling, and B. R. Helliker. 1997. C-4 photosynthesis, atmospheric CO_2 and climate. *Oecologia* **112**(3):285–299.

Evans, R. D. 2001. Physiological mechanisms influencing plant nitrogen isotope composition. *Trends Plant Science* **6**(3):121–126.

Farquhar, G. D., and L. A. Cernusak. 2005. On the isotopic composition of leaf water in the non-steady state. *Functional Plant Biology* **32**(4):293–303.

Farquhar, G. D., J. R. Ehleringer, and K. T. Hubick. 1989. Carbon isotope discrimination and photosynthesis. *Annual Review of Plant Physiology and Plant Molecular Biology* **40**:503–537.

Farquhar, G. D., J. Lloyd, J. A. Taylor, L. B. Flanagan, J. P. Syvertsen, K. T. Hubick, S. C. Wong, and J. R. Ehleringer. 1993. Vegetation effects on the isotope composition of oxygen in atmospheric CO_2. *Nature* **363**(6428):439–443.

Faure, G., and J. L. Powell. 1972. *Strontium Isotope Geology.* Springer-Verlag, New York.

Fekete, B. M., J. J. Gibson, P. Aggarwal, and C. J. Vorosmarty. 2006. Application of isotope tracers in continental scale hydrological modeling. *Journal of Hydrology* **330**:444–456.

Flanagan, L. B., J. P. Comstock, and J. R. Ehleringer. 1991. Comparison of modeled and observed environmental influences on the stable oxygen and hydrogen isotope composition of leaf water in *Phaseolus vulgaris* L. *Plant Physiology* **96**(2):588–596.

Handley, L. L., A. T. Austin, D. Robinson, C. M. Scrimgeour, J. A. Raven, T. H. E. Heaton, S. Schmidt, and G. R. Stewart. 1999. The N-15 natural abundance (delta N-15) of ecosystem samples reflects measures of water availability. *Australian Journal of Plant Physiology* **26**(2):185–199.

Helliker, B. R., and J. R. Ehleringer. 2000. Establishing a grassland signature in veins: O-18 in the leaf water of C-3 and C-4 grasses. *Proceedings of the National Academy of Sciences of the United States of America* **97**(14):7894–7898.

Hemming, D., D. Yakir, P. Ambus, M. Aurela, C. Besson, K. Black, N. Buchmann, R. Burlett, A. Cescatti, R. Clement, P. Gross, A. Granier, *et al.* 2005. Pan-European d13C values of air and organic matter from forest ecosystems. *Global Change Biology* **11**:1065–1093.

Hobson, K. A., and L. I. Wassenaar. 1997. Linking breeding and wintering grounds of neotropical migrant songbirds using stable hydrogen isotopic analysis of feathers. *Oecologia* **109**:142–148.

Hobson, K. A., L. I. Wassenaar, B. Mila, I. Lovette, C. Dingle, and T. B. Smith. 2003. Stable isotopes as indicators of altitudinal distributions and movements in an Ecuadorean hummingbird community. *Oecologia* **136**(2):302–308.

Hobson, K. A., G. J. Bowen, L. I. Wassenaar, Y. Ferrand, and H. Lormee. 2004. Using stable hydrogen and oxygen isotope measurements of feathers to infer geographical origins of migrating European birds. *Oecologia* **141**:477–488.

Hoppe, K. A., P. L. Koch, R. W. Carlson, and S. D. Webb. 1999. Tracking mammoths and mastodons: Reconstruction of migratory behavior using strontium isotope ratios. *Geology* **27**(5):439–442.

Houlton, B. Z., D. M. Sigman, and L. O. Hedin. 2006. Isotopic evidence for large gaseous nitrogen losses from tropical rainforests. *Proceedings of the National Academy of Sciences of the United States of America* **103**(23):8745–8750.

Ingraham, N. L., and B. E. Taylor. 1991. Light stable isotope systematics of large-scale hydrologic regimes in California and Nevada. *Water Resources Research* **27**(1):77–90.

Isaaks, E. H., and R. M. Srivastava. 1990. *An Introduction to Applied Geostatistics.* Oxford University Press, New York.

Kendall, C., and T. B. Coplen. 2001. Distribution of oxygen-18 and deuterium in river waters across the United States. *Hydrological Process* **15**(7):1363–1393.

Kennedy, M. J., O. A. Chadwick, P. M. Vitousek, L. A. Derry, and D. M. Hendricks. 1998. Changing sources of base cations during ecosystem development, Hawaiian Islands. *Geology* **26**(11):1015–1018.

Lloyd, J., and G. D. Farquhar. 1994. ^{13}C discrimination during CO_2 assimilation by the terrestrial biosphere. *Oecologia* **99**(3–4):201–215.

Longinelli, A., and E. Selmo. 2003. Isotopic composition of precipitation in Italy: A first overall map. *Journal of Hydrology* **270**(1–2):75–88.

Luo, Y.-H., and L. Sternberg. 1992. Spatial D/H heterogeneity of leaf water. *Plant Physiology* **99**(1):348–350.

MacFadden, B. J., and T. E. Cerling. 1994. Fossil horses, carbon isotopes and global change. *Tree* **9**:481–485.

Meehan, T. D., J. T. Giermakowski, and P. M. Cryan. 2004. GIS-based model of stable hydrogen isotope ratios in North American growing-season precipitation for use in animal movement studies. *Isotopes In Environmental and Health Studies* **40**(4):291–300.

Murphy, B. P., and D. Bowman. 2006. Kangaroo metabolism does not cause the relationship between bone collagen delta N-15 and water availability. *Functional Ecology* **20**(6):1062–1069.

Murphy, B. P., D. M. J. S. Bowman, and M. K. Gagan. 2007. The interactive effect of temperature and humidity on the oxygen isotope composition of kangaroos. *Functional Ecology* **21**(4):757–766.

Naiman, Z., J. Quade, and P. J. Patchett. 2000. Isotopic evidence for eolian recycling of pedogenic carbonate and variations in carbonate dust sources throughout the southwest United States. *Geochimica et Cosmochimica Acta* **64**(18):3099–3109.

New, M., D. Lister, M. Hulme, and I. Makin. 2002. A high-resolution data set of surface climate over global land areas. *Climate Research* **21**(1):1–25.

Norris, D. R., P. P. Marra, R. Montgomerie, T. K. Kyser, and L. M. Ratcliffe. 2004. Reproductive effort, molting latitude, and feather color in a migratory songbird. *Science* **306**:2249–2250.

O'Leary, M. H. 1981. Carbon isotope fractionation in plants. *Phytochemistry* **20**:553–567.

O'Leary, M. H., S. Madhavan, and P. Paneth. 1992. Physical and chemical basis of carbon isotope fractionation in plants. *Plant Cell Environment* **15**(9):1099–1104.

Ometto, J. P. H. B., J. R. Ehleringer, T. F. Domingues, J. A. Berry, F. Y. Ishida, E. Mazzi, N. Higuchi, L. B. Flanagan, G. B. Nardoto, and L. A. Martinelli. 2006. The stable carbon and nitrogen isotopic composition of vegetation in tropical forests of the Amazon Basin, Brazil. *Biogeochemistry* **79**(1–2):251–274.

Osmond, C. B., W. G. Allaway, B. G. Sutton, J. H. Troughton, O. Queiroz, U. Luttge, and K. Winter. 1973. Carbon isotope discrimination in photosynthesis of CAM plants. *Nature* **246**:41–42.

Pain, D., R. Green, B. Gießing, A. Kozulin, A. Poluda, U. Ottosson, M. Flade, and G. Hilton. 2004. Using stable isotopes to investigate migratory connectivity of the globally threatened aquatic warbler Acrocephalus paludicola. *Oecologia* **138**(2):168–174.

Pardo, L. H., P. H. Templer, C. L. Goodale, S. Duke, P. M. Groffman, M. B. Adams, P. Boeckx, J. Boggs, J. Campbell, B. Colman, J. Compton, B. Emmett, *et al.* 2006. Regional assessment of N saturation using foliar and root delta N-15. *Biogeochemistry* **80**(2):143–171.

Park, R., and S. Epstein. 1960. Carbon isotope fractionation during photosynthesis. *Geochimica Cosmochimica Acta* **21**:110–126.

Pataki, D. E., D. R. Bowling, and J. R. Ehleringer. 2003. Seasonal cycle of carbon dioxide and its isotopic composition in an urban atmosphere: Anthropogenic and biogenic effects. *Journal of Geophysical Research-Atmospheres* **108**(D23):4735.

Peters, C. R., and J. C. Vogel. 2005. Africa's wild C-4 plant foods and possible early hominid diets. *Journal of Human Evolution* **48**(3):219–236.

Peterson, T. C., and R. S. Vose. 1997. An overview of the Global Historical Climatology Network temperature data base. *Bulletin of the American Meteorological Society* **78**:2837–2849.

Quade, J., A. R. Chivas, and M. T. McCulloch. 1995. Strontium and carbon isotope tracers and the origins of soil carbonate in South Australia and Victoria. *Palaeogeography Palaeoclimatology Palaeoecology* **113**:103–117.

Robinson, D. 2001. $\delta^{15}N$ as an integrator of the nitrogen cycle. *Tree* **16**(3):153–162.

Roden, J. S., G. G. Lin, and J. R. Ehleringer. 2000. A mechanistic model for interpretation of hydrogen and oxygen isotope ratios in tree-ring cellulose. *Geochimica et Cosmochimica Acta* **64**(1):21–35.

Rozanski, K., L. Araguas-Araguas, and R. Gonfiantini. 1993. Isotopic patterns in modern global precipitation. Pages 1–36 *in* P. K. Swart, K. C. Lohmann, J. McKenzie, and S. Savin (Eds.), *Climate Change in Continental Isotopic Records.* American Geophysical Union, Washington, DC.

Scholze, M., J. O. Kaplan, W. Knorr, and M. Heimann. 2003. Climate and interannual variability of the atmosphere-biosphere 13 CO. *Geophysical Research Letters* **30**(2):1097.

Sponheimer, M., and J. A. Lee-Thorp. 1999. Isotopic evidence for the diet of an early hominid, Australopithecus africanus. *Science* **283**(5400):368–370.

Sternberg, L., and M. J. Deniro. 1983. Isotopic composition of cellulose from C3, C4, and CAM plants growing near one another. *Science* **220**(4600):947–949.

Still, C. J., J. A. Berry, G. J. Collatz, and R. S. DeFries. 2003. Global distribution of C_3 and C_4 vegetation: Carbon cycle implications. *Global Biogeochem Cycle* **17**:1006.

Suits, N. S., A. S. Denning, J. A. Berry, C. J. Still, J. Kaduk, J. B. Miller, and I. T. Baker. 2005. Simulation of carbon isotope discrimination of the terrestrial biosphere. *Global Biogeochem Cycle* **19**:GB1017.

Trusdell, F. A., E. W. Wolfe, and J. Morris. 2006. Digital Database of the Geological Map of the Island of Hawaii. Data Series. US Geological Survey, http://pubs.usgs.gov/ds/2005/144/.

Van Der Merwe, N. J., J. A. L. Throp, and R. H. V. Bell. 1988. Carbon isotopes as indicators of elephant diets and African environments. *Biological Abstracts* [Vol. 86, Iss. 8, Ref. 78664]. **26**(2):163–172.

Von Caemmerer, S. 1992. Carbon isotope discrimination in C3-C4 intermediates. *Plant Cell Environment* **15**(9):1063–1072.

Wang, L. X., G. S. Okin, J. Wang, H. Epstein, and S. A. Macko. 2007. Predicting leaf and canopy N-15 compositions from reflectance spectra. *Geophysical Research Letters* **34**: Art No. L02401.

Wassenaar, L. I., and K. A. Hobson. 2001. A stable-isotope approach to delineate geographical catchment areas of avian migration monitoring stations in North America. *Environmental Science & Technology* **35**(9):1845–1850.

West, J. B., J. HilleRisLambers, T. D. Lee, S. E. Hobbie, and P. B. Reich. 2005. Legume species identity and soil nitrogen supply determine symbiotic nitrogen-fixation responses to elevated atmospheric [CO_2]. *New Phytologist* **167**(2):523–530.

West, J. B., J. R. Ehleringer, and T. E. Cerling. 2007. Geography and vintage predicted by a novel GIS model of wine $\delta^{18}O$. *Journal of Agricultural and Food Chemistry* **55**(17):7075–7083.

West, J. B., A. Sobek, and J. R. Ehleringer. A simplified GIS approach to modeling global leaf water isoscapes. *Global Ecology and Biogeography*, in review.

Wickman, F. E. 1952. Variations in the relative abundance of the carbon isotopes in plants. *Geochimica Cosmochimica Acta* **2**:243–254.

Widory, D. 2007. Nitrogen isotopes: Tracers of origin and processes affecting PM10 in the atmosphere of Paris. *Atmospheric Environment* **41**(11):2382–2390.

Wu, C. F. J. 1986. Jackknife, bootstrap and other resampling methods in regression analysis. *Annals of Statistics* **14**(4):1261–1295.

Wunder, M. B., C. L. Kester, F. L. Knopf, and R. O. Rye. 2005. A test of geographic assignment using isotope tracers in feathers of known origin. *Oecologia* **144**:607–617.

Yakir, D., M. J. DeNiro, and J. R. Gat. 1990. Natural deuterium and oxygen-18 enrichment in leaf water of cotton plants grown under wet and dry conditions: Evidence for water compartmentation and its dynamics. *Plant Cell Environment* **13**(1):49–56.

Yakir, D., J. A. Berry, L. Giles, and C. B. Osmond. 1994. Isotopic heterogeneity of water in transpiring leaves: Identification of the component that controls the 18O of atmospheric O_2 and CO_2. *Plant Cell Environment* **17**:73–80.

Yurtsever, Y., and J. R. Gat. 1981. Atmospheric waters. Page 103–142 *in.* J. R. Gat and R. Gonfiantini (Eds.), *Stable Isotope Hydrology: Deuterium and Oxygen-18 in the Water Cycle.* International Atomic Energy Agency, Vienna.

CHAPTER 5

Analysis and Design for Isotope-Based Studies of Migratory Animals

Michael B. Wunder* and D. Ryan Norris[†]

*Department of Fish, Wildlife, and Conservation Biology, Colorado State University
[†]Department of Integrative Biology, University of Guelph

Contents

I. INTRODUCTION

One of the major advantages of using stable isotopes as intrinsic markers is that migratory individuals only have to be captured once to estimate a geographic move. However, the problem of geographic *assignments* (Table 5.1) to individuals of unknown origin also presents some significant quantitative

Tracking Animal Migration with Stable Isotopes
K. A. Hobson and L. I. Wassenaar (Editors)
ISSN 1936-7961, DOI: 10.1016/S1936-7961(07)00005-X

TABLE 5.1	Glossary of key terms associated with tracking animals using stable isotopes
Term	**Definitions in the context of isotope-based assignments**
Assignment method	A way to place animals of unknown origin in a specific location during a previous period of the migratory cycle
Assign-time calibration	The use of known-location tissues to relate isotope values to geography and environmental material
Asymptotic standard deviation	The standard deviation that applies to very large sample sizes. This is not the value that applies to a single IRMS analysis run. It would apply if we were able to repeat our samples runs many times.
Bayes' Rule	A formal rule for inverting conditional probabilities: $P(A|B) = P(B|A) * P(A)/P(B)$
Deterministic	Refers to an outcome that is not influenced by chance events. The structure of the process is known and well-defined.
Isoline	A line across space that represents equal isotopic values
Isoscape or base map	A map of isotopic values used to assign the locations of animals of unknown origin
Likelihood	The hypothetical probability an event in the past would have resulted in a particular outcome
Mean field	The asymptotic result for a stochastic process; the set of average values for a model
Monte Carlo integration	A simulation-based method for approximating a probability distribution function
Probability	The chance that a specified event will occur in the future
Run-time calibration	The use of known-value material (standards) to relate IRMS measurements to true values. The term run-time indicates that an independent calibration is made for each carosel of samples run using continuous flow IRMS
Stochastic	Refers to an outcome that depends on chance, a random event
Spatial interpolation	A method to generate predicated values for all points in space from a finite number of data points

and statistical challenges. This chapter reviews and discusses quantitative approaches for making inferences about the geographic history of migratory animals using stable isotope values measured in sampled tissues. The most commonly used tissues include feathers, claws, fur, blood, muscle, or bone (see Chapter 2). Here, we describe the nature of stable isotope data and its direct relevance to estimating the origin of individuals, discuss some common assumptions, and show how the basic approach to assignment can be treated as a *calibration* problem. We then review and analyze modeling approaches that have been used to date (Table 5.2), briefly discuss potential for future extensions and improvements, and conclude with a discussion of sampling design considerations. Our goal in this chapter is to provide professionals and graduate students a comprehensive introduction to methods that have been used to assign individuals of unknown origin using stable isotopes.

II. CALIBRATION AND THE ASSIGNMENT PROBLEM

An essential premise in using stable isotopes to track the movement of migratory animals is that detectable and predictable patterns (or at least differences) exist in the spatial distributions of stable isotopes in the environment from where migratory organisms obtained their diet (Chapter 3). For example, as noted in earlier chapters, there are relatively strong, well-known, and globally predictable geographic gradients in the hydrogen (δD) and oxygen ($\delta^{18}O$) isotope compositions of meteoric water (Dansgaard 1964). Geographic gradients in other isotopes are less well known. Stable strontium isotopes ($\delta^{87}Sr$) vary with the age and type of bedrock (Chapter 4). Spatial distribution models of

$\delta^{13}C$ depend strongly on modeled distributions of C3 and C4 plants as well as models of plant respiration and photosynthesis and potentially of agricultural crops. We are unaware of any biogeophysical model or theory for describing expected spatial distribution of $\delta^{15}N$ for any organisms' tissues.

When geographic patterns are detected for stable isotopes, they are really inferred from a finite number of sampling points across the landscape. Although patterns in δD of feathers have been modeled as linear functions of latitude and longitude (Kelly *et al.* 2002, Rubenstein *et al.* 2002), a *spatially interpolated model* (Table 5.1) is necessary to derive a continuous surface over which animals can be assigned to specific locations. For example, δD values in precipitation have been empirically modeled for North America (Hobson and Wassenaar 1997, Meehan *et al.* 2004) and worldwide (Bowen *et al.* 2005). The generation of these models produces *isoscapes* (Table 5.1) in which individuals of unknown origin can be assigned (Chapter 4).

The key assumption, however, is that patterns in stable isotopes derived from environmental or low trophic-level sources are faithfully maintained or translated through food webs. In other words, the *isotopic discrimination* (Table 5.1) between the stable isotopes in dietary sources and the tissue in the species of interest is both predictable and constant in time and space (Chapter 3). Thus, an animal incorporates isotopic values that are assumed to be representative of the location at which the tissue was grown. When the animal moves elsewhere, it can then be sampled to infer its previous geographic location. Quantitatively, then, in order to infer the geographic location most likely associated with measured stable isotope values, the most critical part of the approach is to calibrate the geographic model using tissues from animals of known origin. When the model is in explicit geographic map form, the resultant calibrated maps are often referred to as isotopic base maps.

Just as all isotope-ratio mass spectrometry data depend on *run-time calibrations* (Table 5.1) with universally accepted reference materials, or calibrated standards, so too do models for assigning individuals to locations depend critically on *assign-time calibrations* (Table 5.1) using "standards" (tissues of known geographic origin). The accuracy of isotope ratio mass spectrometer (IRMS) measurements is best when using standards with chemical compositions similar to that of the unknown samples (*e.g.*, using organic keratin standards to calibrate organic keratin samples; Wassenaar and Hobson 2002). In the same way, the accuracy of geographic model calibrations is improved by using "standards" with similar attributes as the samples for which we want to assign geographic provenance. This means that geographic assignment models perform best when calibrated using standards from the same species as the unknown samples, and that bracket the range of isotope values and all other covariates in both time and space from which the samples are drawn.

Ideally then, the calibration dataset would include samples obtained from across all potential areas in which the migratory species of interest could have originated. However, sampling tissues from all potential places of origin is not only costly, but very often logistically difficult or impossible for areas well beyond the reach of monitoring and sampling networks. Most studies have instead simply used previously estimated δD isoscapes, with the assumption that the predicted gradients in δD of precipitation are directly transferred to bird feathers (*e.g.*, Wassenaar and Hobson 2001, Norris *et al.* 2004, Bearhop *et al.* 2005, Mazerolle and Hobson 2005, Mazerolle *et al.* 2005, Boulet *et al.* 2006, Hobson *et al.* 2006, Norris *et al.* 2006, Perez and Hobson 2006, Hobson *et al.* 2007, Mazerolle and Hobson 2007). In mathematical terms, this assumption implies a slope of one and a fixed intercept (usually assumed to range from $-20‰$ to $-25‰$) for the regression of the isotope values in feathers on those for precipitation (Table 5.1, Chapter 3). The derivation of the intercept is typically done either by: (1) adopting a discrimination factor estimated from another species in a previous study (most common) or (2) estimating a discrimination factor for the species of interest but only from a single or few location(s). However, there have been relatively few studies that have addressed the assumption of a constant discrimination rate across space and between species (Lott *et al.* 2003).

Alternatively, isoscapes can be derived directly from tissues sampled from the species of interest over the time period of interest (Hobson *et al.* 1999). In such cases, the calibration is incorporated directly into the spatial interpolation. Once developed, an isoscape is a static surface that is generally then

TABLE 5.2 Description of methods for assigning animals of unknown origin using stable isotopes

Level of complexity	Model type	Description	Advantages	Disadvantages	Incorporates sources of error	Key references
Low	Map lookup	Isotope value of animal assigned to area based on isolines or isoscapes on a map	Easy to implement	Not rooted in probability, does not incorporate spatial variability of isotopes or other known source of error	No	Chamberlain et al. (1997), Hobson and Wassenaar (1997), Meehan et al. (2001), Bearhop et al. (2005), Lott and Smith (2006), Hobson et al. (2007), Paxton et al. (2007)
	Linear regression	Origin inferred based on a regression of isotopes on latitude and/or longitude	Easy to implement	Does not incorporate spatial variability of isotopes within regions or other error sources	No	Kelly et al. (2002), Rubenstein et al. (2002)
	Classification trees	Origin inferred based on series of hierarchical, discrimination-based decision rules	Can be applied to multiple isotopes, does not require distributional assumptions, accommodates both continuous and categorical predictors	Does not incorporate most error, no a priori hierarchy for multiple isotopes, does not provide degree of certainty for branching decisions	No	Hebert and Wassenaar (2005a,b)
	Likelihood-based assignments	Origin inferred from probability density functions for isotope values from given regions	Can be applied to multiple isotopes, provides probability of assignment for given individual, easy to implement	Does not incorporate most error, simply assigns region for animal based on highest likelihood value, requires sampling all potential regions of origin	Some	Royle and Rubenstein (2004), Kelly et al. (2005), Wunder et al. (2005), Norris et al. (2006)

Likelihood with priors	Same as above but adds prior information (isotope or other sources using Bayes' Rule)	Can be applied to multiple isotopes can utilize non-isotopic information (see Table 5.3)	Does not incorporate most error, simply assigns region for animal based on highest posterior probability, requires samplings all potential of origin	Some	Royle and Rubenstein (2004), Wunder et al. (2005), Norris et al. (2006)
Stochastic extension of likelihood	Same as above but adds known sources of error associated with isotope data	Can be applied to multiple isotopes, incorporates multiple sources of error, provides a range of possible assignments for a given animal	Computationally intensive	Yes	Wunder and Norris (2008)
Probability surfaces	Models stochastic error process over mean field surface	As above, incorporates known variance sources is applied as a continuous probability surface over space	Computationally intensive	Yes	Wunder (2007)

High →

treated as a *deterministic* (Table 5.1) process to describe the spatial pattern in the isotope of interest. As we discuss below, all these calibrated models represent only the first step in the modeling process for assigning provenance to migratory animals. Ultimately, we would like to get beyond these simple cartographic exercises to models that provide surfaces for describing geographically explicit probabilities of origin for individual animals.

Recent studies have demonstrated that there can be substantial variation in isotope values (especially δD) in feathers from a group of individuals sampled at a given site, or even within a single tissue type (Chapter 2). Isotopic differences between age groups represent one of the best-documented examples (Duxbury *et al.* 2003, Meehan *et al.* 2003, Smith and Dufty 2005, Langin *et al.* 2007, Wunder 2007). Lott *et al.* (2003) demonstrated that there are geographic differences in the relationship between δD in feathers and precipitation, and Wunder *et al.* (2005) documented within-location differences among years for δD, $\delta^{13}C$, and $\delta^{15}N$ in mountain plovers (*Charadrius montanus*). Langin *et al.* (2007) showed systematic differences in δD between blood and feathers between known-origin American redstart (*Setophaga ruticilla*) adults and nestlings. These findings emphasize the need for more studies aimed at understanding the mechanisms that drive differences in isotopic discrimination factors over space, between species, and across life histories.

Hobson (2005) noted the utility of developing isoscapes that are derived directly from the tissues of interest (*e.g.*, bird feathers), but only one such effort has been undertaken to date (Hobson *et al.* 1999). Such endeavors require broad spatial sampling for the species (and tissue) of interest and require that multiple individuals be sampled at each location to estimate within-site variation. Another need for localized calibration arises because broad geographic patterns most likely change over time (*e.g.*, they vary seasonally, annually, by decade, by century). Thus, it is best to calibrate models with known origin tissues collected from a representative time period of interest. For example, assigning wintering migratory birds to breeding origins would be optimized by sampling known-origin tissues on the breeding grounds in the previous year. And because the relationship between broad geospatial patterns likely varies with species, age of and condition of individuals as well as with a host of environmental factors, calibrations with known origin tissues should ideally cover the expected range for all of these covariates. In other words, the most robust models for estimating the origin of migratory animals are those that are calibrated as specifically as possible, both in space and over an appropriate time frame.

We recognize that in most cases, developing species- and tissue-specific isoscapes may be unrealistic due to both logistical difficulties (*e.g.*, inaccessible areas) and financial constraints. There is, therefore, a critical need to test general hypotheses that attempt to explain observed differences in isotope discrimination factors between and within species, between different tissues, and across space. Testing should include a combination of *in situ* studies of wild animals and controlled laboratory studies, ideally in a back-and-forth way that promotes understanding of mechanistic factors in natural settings. Only by advancing understanding in this way we will be able to derive some general rules about variation in isotope discrimination and be able to identify optimal sampling schemes for generating more robust isoscapes that can be used to track the movement of many migratory species.

III. CHARACTERISTICS OF ISOTOPE DATA FOR USE IN STUDYING MIGRATORY ANIMALS

Measurements of stable isotope values using IRMS are reported relative to a known internationally accepted reference and use the typical "δ" notion (Chapter 2). The δ-values are ratios of ratios that can range from $-1000‰$ to positive infinity. The δ-values are calculated using a calibration that usually consists of a two-point (or a few-point) linear regression using reference materials (standards) that (should) span the natural isotopic range of the unknown samples. The full range of possible values is never covered by any natural dataset, and the range of interest is almost always narrow enough that a careful linear calibration is an excellent approximation (Gröning 2004). The important point to recognize is that the δ-value of a sample is not an exact value, but is derived from a linear regression of results from primary reference materials (standards).

Jardine and Cunjak (2005) remind us that IRMS measurement of bulk unknown samples (of variable mass and isotopic heterogeneity) can not be 100% accurate and are affected by inherent variability in the preparative procedures. Although the IRMS apparatus itself is capable of very fine precision (*e.g.*, calculating ratios to the sixth decimal place), the magnitude of the δ differences we may want to detect in the environment can be on the same order as the precision of the instrumentation. This point should not be overlooked by those wanting to use stable isotopes in ecology. It is not reasonable, for example, to claim that a difference of 10‰ between two populations is significant if the analytical error associated with the particular isotope measurements is ±3‰. This is because the ANOVA or other similar tests for significant differences between means assume that the δ-values are *exactly* known. However, the δ-values are each only known within ±3‰, which means that it is not unreasonable to expect to observe two different measurements for the same sample (not two samples from the same population) that are 6‰ apart. More importantly, this suggests the utility of using *stochastic* (Table 5.1) approaches to analyze conclusions about stable isotope data.

IV. STATISTICAL ASSUMPTIONS FOR USING ISOTOPE DATA TO INFER ORIGIN

A. Statistical Independence

The assumption of statistical independence requires samples be drawn at random from the at-large target population of interest. In linear models that relate stable isotope values to latitude, for example, clustered sampling designs (*e.g.*, many individuals from a few locations) illustrate a classic violation of this assumption. For example, suppose that feathers from 140 birds were sampled for δD. Suppose that those 140 birds were taken from only 10 sites that vary in latitude. In this case, there are not really 138 degrees of freedom for fitting a linear model (140 minus 1 for the slope and 1 for the intercept); there are only 8 degrees of freedom. In this example, site is the more appropriate sample unit because we are trying to relate stable isotope values to geographic variables, and each of the 14 birds sampled from a single site shares the same value for the response variable (latitude). The samples for each site are better treated as replicates than as independent samples. Statistically, this problem can be overcome by including a random effects term (site) in the linear model, or considering a repeated measures modeling framework.

B. Identically Distributed (Process Homogeneity)

Implicit in all stable isotope studies that seek provenance of migratory animals is the assumption that all individuals in the population derived from a particular location are subjected to the same processes that generate variance in tissue isotope values. For example, we assume that all individuals of the target species respond in the same way to environmental stresses, foraging at roughly the same position in relatively similar food webs, and developing tissues at roughly the same rate. This strongest assumption that a site (characterized by the measurement of multiple individuals) is isotopically homogeneous will almost never be met.

Isotopic variation among individuals stems from differences in what they consume, when they consume it, and under what conditions they develop tissues, all of which potentially contain useful information. The second source of stable isotopic variability comes from isotopic heterogeneity within an individual animal. Organic tissues within an animal may develop and senesce at different rates. As such, this too can provide useful information about changes in geography, environmental conditions, diet, or life history trade-offs. The third source of variance is IRMS measurement error derived from run-time calibrations, and it is a simple matter to measure the calibration residuals to quantify analytical error. Although the error is usually not the same for each analysis run, the *asymptotic standard deviation* (Table 5.1) of these residuals is generally all that is reported. As more is learned about systematic deviations from this assumption of process homogeneity, relevant covariates can be added to adjust models.

V. POPULATION-LEVEL VERSUS INDIVIDUAL-LEVEL GEOGRAPHIC ASSIGNMENTS

Are we interested in the distribution of geographic origins estimated from δ-values for individuals in a sample population of migratory animals, or are we interested in the geographic origin associated with the average δ-value for a population of migratory animals? These subtly different questions have traditionally been treated in different ways. However, we argue that population-level questions about the origin of migratory animals are best treated by compiling individual-level assignments rather than summarizing δ-values from a given sample of individuals (*e.g.*, using the average δ-value). This is because we are not really interested in δ-values themselves but rather in a transformation of those values to some geographic location(s). Because there is not a perfect 1:1 transformation from δ-value to geographic coordinates, it is important to transform the δ-value for each individual data point into a geographic value *before* determining the population-level characteristics of the distribution of geographies. Otherwise, we may potentially lose very valuable information associated with sources of "error," or variation that may be systematic, and therefore informative.

Assigning individuals to geographic origins one at a time and exploring properties of the resultant geographic distribution is more flexible and natural than using the average of the δ-values to determine the average geographic origin for a sample population for two important reasons: (1) it does not force the data to follow the normal distribution (or any other assumed distribution) and (2) it does not provide an overly optimistic sense of precision by (a) ignoring among- or within-individual variance in the transformation to geography and (b) ignoring measurement error. More importantly, the goal of many, if not most studies of migratory animals, is to determine if there is any geographic structure in the sampled population. In other words, the goal is to describe the distribution of assignments for the sampled population. It is therefore counterproductive to fit an assumed distribution (structure) for the population prior to doing the assignment analysis.

First, describing a population by calculating the arithmetic average (mean) and standard deviation for the isotopic data implicitly assumes normality. The mean (μ) and standard deviation (σ) are the two parameters that fully specify the normal distribution; μ is the location and σ is the scale. However, if the sample population is actually a mix of two distinct populations, it should theoretically follow a bimodal distribution. Using μ and σ to describe the distribution will put most of the probability mass in a location between the two modes, where few data actually exist. If we leave the distribution alone as bimodal, the bulk of the probability mass will be defined by two disjunct ranges.

As an example, let us consider a case where we wish to determine the North American breeding origins for a population of migratory songbirds wintering in the Caribbean. To do this, we might calculate a mean and standard deviation of δD values from feathers of birds at the wintering population. Because these feathers were grown on the breeding grounds, we can then infer the origin of this population based on a δD isoscape derived for North America. To do this, we could calculate the mean and standard deviation of the δD values for the entire sample population and then shade the area of the δD isoscape that is centered on the mean and extends to either side by one standard deviation or some other scaling amount. Most of the shaded area of our map might, for example, be somewhere around New Brunswick and Maine. However, on closer inspection, the δD values from individuals sampled at the wintering population indicate an apparent bimodal distribution. If we were to assign individuals to breeding origins first, we would see that the population actually originated from one of two places: either Kentucky or Newfoundland but nowhere in between. Thus, it is important to first assign individuals to places of origin and then summarize this information based on the actual distribution of values that is generated by the population rather than assume a normal distribution and assign the population based on a mean and standard deviation.

The second reason for first transforming the δ-values to geographic values for each individual is less obvious. The δ-values in the same tissue from different individuals from the same population from the same single location are never identical. Repeated samples from the same individual are not even expected to be identical. Because of this, the transformation from δ-values to geographic location is not done with complete certainty. By assigning individuals to geographic locations first, we propagate the uncertainty associated with the transformation and provide a less biased answer about the geographic structure of the sample population.

For these two primary reasons, we suggest that the more conservative and potentially more informative approach is to consider the assignment of origins (transformation of δ-values to geographic locations) for each individual and then use the resulting distribution of geographic origins to address population-level questions. In Section VI, we show that only some of the assignment methods used to date can accommodate this perspective.

VI. MODELING APPROACHES

A. Map Lookup

The map-lookup approach is very straightforward. What we call map lookup simply involves defining geographic gradients (base maps) of isotope values, measuring an individual of unknown origin, and then assigning it to the area of the mapped gradient that corresponds to its isotope value. Typically, this approach capitalizes on the spatial patterns generated from indirect sources of information (*i.e.*, rainfall, primary producers) that are used to generate isoscapes.

The studies that originally jumpstarted the use of stable isotopes to study migratory movements (Chamberlain *et al.* 1997, Hobson and Wassenaar 1997) used the map lookup approach to illustrate the utility of δD for determining the provenance of migratory birds and it is still among the most commonly employed (Wassenaar and Hobson 2001, Norris *et al.* 2004, Bearhop *et al.* 2005, Mazerolle and Hobson 2005, Mazerolle *et al.* 2005, Boulet *et al.* 2006, Hobson *et al.* 2006, Lott and Smith 2006, Hobson *et al.* 2007, Paxton *et al.* 2007). The appeal of this approach is that it is easy to understand and apply because assignments of unknown animals require no additional computation.

Map lookup approaches treat the output from spatial interpolation models (isoscapes) as a known pattern-generating process. In other words, the baseline assumption is that the modeled pattern for, say, δD in precipitation effectively mimics the "truth." The task for any particular study then becomes to simply calibrate the map. In the case of δD, this calibration is an estimate of the isotope discrimination factor between rainfall and a tissue in the species of interest.

In practice, this approach has been limited to δD, and most studies focus on birds, usually sampling keratin in feathers or claws. For this reason, we limit our discussion here to studies of migratory birds using δD in keratin. The best performance for the map lookup approach will be obtained when researchers are calibrating isoscapes (base maps) with known origin tissues that are the same tissue type from the same species, age class, and habitats as the samples of interest; samples that were collected from across the full extent of the geographic range and that were collected during the same years as the samples obtained from migratory individuals of unknown origin. In the vast majority of cases, however, these data are simply not available. If researchers cannot provide study- or species-specific calibrations, the next best performance model will be obtained from published calibrations that correspond most closely to the study species and geographic range. Obtaining specific calibrations, however, is not only a recommended approach for the map lookup method but for all methods used for assignments.

There are some broad consistencies in the observed relationship between δD in precipitation and that of feather keratin within some geographic regions and among some species groups that can be used as crude calibration guidelines, but we note that published results vary quite a bit (see Table 3.1 in Chapter 3). For example, Hobson and Wassenaar (1997) estimated a single calibration curve for a collection of six species of songbirds that breed in the forests of eastern North America as $y = x - 34‰$, where y is δD for feathers and x is modeled δD for precipitation; this finding suggests a 1:1 transformation of isoscape precipitation values to expected feather values (*i.e.*, slope $= 1$). Hobson *et al.* (2001) estimated the same calibration curve for a different passerine of the eastern North American forest, the Bicknell's Thrush (*Catharus bicknelli*), as $y = 0.68x - 26.1‰$. The slope of the model (0.68) implies more than 30% expansion of modeled precipitation values compared with feather values. This suggests broad within-site variation relative to modeled precipitation values. Meehan *et al.* (2004) offer four calibrations for relating feathers to an elevation-corrected isoscape they present. Slopes range from 0.6 to 0.9, including a calibration of $y = 0.7x - 21‰$ for Wilson's warbler, the same species for which Paxton *et al.* (2007) later estimated $y = 1.4x + 14.467‰$ using the same isoscape but a sample collected at a different time and from different locations. In the first case, there is a 30% expansion of modeled δD in precipitation relative to feathers. In the second case, there is a 40% compression. This strongly suggests that it is unreliable to use a single calibration function for the Meehan *et al.* (2004) isoscape and that *case*-specific calibration will produce the most robust models.

North American Accipiter hawks have been well studied: Meehan *et al.* (2001) estimated a calibration between Cooper's hawk (*Accipiter cooperi*) feathers and the Hobson and Wassenaar (1997) isoscape as $y = x - 34‰$, implying the same 1:1 mapping as for some eastern North American songbirds. Using the same isoscape, Lott *et al.* (2003) estimated six different calibrations for a collection of nine species of diurnal North American raptors. The six groupings were based on diet and whether the species foraged in coastal areas. The associated calibration slopes ranged from -0.59 to 0.62, interestingly ranging from negative to positive associations between feathers and model output. Bowen *et al.* (2005) offer a different global isoscape along with one calibration specific to North American songbirds ($y = 1.07x - 19.4‰$) and one to European songbirds ($y = 0.85x - 22.3‰$).

Once a calibrated base map is obtained, the question then shifts to "How should we assign our δD values to regions on the map?" There are two distinct ways to represent geographic assignments using the map lookup approach. The first and most common way is to arbitrarily divide the isoscape using discrete *isolines* (Table 5.1), ranges, or "bins" (*e.g.*, Hobson and Wassenaar 1997, Boulet *et al.* 2006, Paxton *et al.* 2007). Animals are then assigned to the line or into the bin that contains the isotope value measured in its tissue. Using this approach, geographic origin for a sampled population can be represented as a histogram showing the number of individuals assigned to each bin. The second approach determines a range of values that "buffer" the value of interest that can be the δ-value for an individual (*e.g.*, Lott and Smith 2006) or the mean δ-value for a population (*e.g.*, Hobson *et al.* 2007). The next step then finds all cells in the isoscape that fall within that range. The buffer extent is usually related to the estimated measurement error or some similar standard deviation-like scaling. We suggest that researchers look to their calibration regressions for guidance in determining the buffer extent. For example, Meehan *et al.* (2001) regressed feather δD against precipitation δD and estimated the 95% confidence interval for *mean* values at $\pm 3‰$, implying a range of 6‰ for *population* means; the 95% *prediction* interval, however, was $\pm 16‰$, implying a range of 32‰ for *individual* δD values.

Hobson *et al.* (2007) used the map lookup method to determine general provenance for sampled populations of three species of birds passing through a migration-monitoring station. Here, the approach was not to look at the collective result of assigning birds to the map individually as in Lott and Smith (2006). Rather, they treated the groups of birds of each species as distinct populations, all of which were assumed to be normally distributed in δD. For each species, they then defined a range of δD values that would theoretically encompass 50% and 75% of a future sample dataset. These ranges were referred to as "tolerance limits." Tolerance limits are calculated as the mean $\pm$ a scaled standard deviation. The standard deviation was scaled by the z score and the χ^2 value associated with the specific sample size (with $\alpha = 0.05$). It is important to realize that these ranges were defined from the sample of δD values, independent of any geographic information. They assumed the Meehan *et al.* (2004) isoscape could be adjusted by a fixed offset of $-25‰$ for all species. All cells in the isoscape that were within the ranges defined above were assigned a value of "1" and all others were assigned a "0."

Tolerance limits were developed for engineering applications to determine the reliability of industrial production processes. In such applications, there is no interest in the structure of the sample population, only in the consistency of the production process; the engineers usually want to know how precise their processes need to be in order to still perform the task for which they were designed. In contrast, the question of interest in most studies of connectivity in migratory animals is related to characterizing the distribution of sampled population of unknown origin. For this reason, calculating the range of δD around the sample mean and then looking up that range on a base map is less informative for studying the geographic structure of the sample than would be the case if birds were looked up individually first.

Lott and Smith (2006) used map lookup for assigning individuals by buffering the δD value for each individual bird by $\pm 8‰$, and then looking up that range on an isoscape. As with all map lookup approaches, cells with values that fell within the range were assigned a value of "1" and cells that fell outside that range were given a value of "0." Thus, their approach did not make discrete categories *a priori* from the continuous isoscape as in other studies. Rather, the isoscape was made discrete by transforming it into a map of ones and zeros for each individual and then summing over these maps. This geographically indexed frequency distribution was then used to describe the structure of the distribution of probable origins for migrants.

In all cases mentioned above, the measured δD values for feathers were mapped directly to an isoscape (using Meehan *et al.* 2004 for all 2007 studies and Hobson and Wassenaar 1997 for all previous studies). Geographic regions for assignment are always characterized as being isotopically exact. In other words, the map lookup approach always results in a map of zeros (not the area of origin) and ones (the area of origin). The map lookup approach, therefore, does not allow direct probabilistic statements about provenance. Because of this, the approach tends to downplay the effects of various sources of variability and error. For example, any predefined region of potential origin will encompass a range of possible values, no matter if that region is circumscribed by isolines in an isoscape (*e.g.*, Hobson and Wassenaar 1997) or by knowledge of the breeding range (*e.g.*, Norris *et al.* 2006). As mentioned above, there are also known sources of error associated with isotope measurements themselves and with the isotopic discrimination between environmental samples and tissues. Only by formally quantifying all these sources of variation can we make statements about the probability that an individual derived from one region versus another. For these reasons, we feel that the field needs to move beyond a map lookup approach and we advocate more advanced methods for estimating the origin of migratory animals.

B. Regression

This is the most simplistic statistical approach for determining the origin of migratory animals. Here, the calibration between tissue values and geography is direct. There is no intermediate model based on precipitation. This method makes the simple assumption of a direct linear relationship between latitude and/or longitude and stable isotope values observed in tissues. The idea is to simply fit a regression line, invert it, and use isotope values measured in feathers to predict the mean latitude or longitude associated with that value. It has not been widely used, primarily because there does not appear to be widespread evidence for a monotonic relationship between latitude/longitude and naturally occurring distributions of isotopes. Because of this, we do not advocate using regression models for estimating the geographic origin of migratory animals. However, regression models can be helpful for exploratory work so long as appropriate caveats are discussed. For example, direct low-order relationships between isotope values and latitude (or altitude) might allow crude differentiation for relative latitudes or altitudes of origin. However, to draw a firm conclusion about such a pattern, it is important to be able to rule out alternative mechanisms for the observed structure, such as within-location differences in diet, age, species, metabolism.

At least three studies relied primarily on regression for data analysis. Kelly *et al.* (2002) related latitude and δD in feathers of Wilson's warblers using linear regression. They used this relationship as indirect evidence for leapfrog migration in the species. Smith *et al.* (2003) used quadratic regression of δD in feathers on capture date during migration to infer that sharp-shinned hawk (*Accipiter striatus*) employ a chain migration pattern. Rubenstein *et al.* (2002) went a step further and used linear regression to predict the latitude and longitude of wintering black-throated blue warblers (*Dendroica caerulescens*) by considering latitude and longitude as response variables that depended on δD and δ^{13}C values in feathers as predictor variables. This is really the only study that has used regression to directly infer latitude. It is worth noting that their model for determining the latitude of origin for a feather (latitude $= -0.097\delta D_f - 1.95\delta^{13}C_f - 12.55$) explained less than half (44%) of the variation in the data, and that the interval for predicting the origin of an individual was as wide as the sampled latitudinal range!

As intuitive as this method might appear to be, we do not advocate this approach except for the most basic exploratory or pilot studies, primarily because inverting the relationship between latitude/longitude and isotope values is not valid. For example, variation in δD is derived from changes in temperature (realized through variation in latitude), distance from an ocean (longitude), and elevation (Dansgaard 1964). The inverse is not true: variation in geographic location is not derived from variation in δD. Inverting a linear relationship like this is statistically invalid because the error term is generally in the y-direction and no error can exist in the x values. When the response variable (y) is latitude, the only way to generate variation in that direction is by sampling different latitudes and within-site variation is error expressed in the x-direction. This problem can be fixed by adding a random effects term for sample location. However, such a solution renders the regression model useless for determining the origin of unknown samples because the sample location then needs to be known *a priori* (Wunder *et al.* 2005).

As we also noted for the map lookup approach, regression models are unable to incorporate multiple known sources error associated with using stable isotopes for estimating the origin of animals. Thus, regression models do not provide a probability of origin for a given area but rather a single deterministic assignment based on the regression line. Given that there are known inherent sources of variability that are difficult to accommodate in regression and map lookup approaches, we describe in the following sections how estimation of origin can be treated as a more formal probabilistic problem.

C. Assignment Methods

Probabilistic assignment methods include a broad range of computational approaches, but all feature a discrete (categorical) response variable. Assignment methods consist of predefining all possible geographic locations of origin, then calibrating an assignment model by characterizing each of those locations with the distribution of stable isotope data. Ideally the characterization is derived from isotopic measurements for individuals known to have grown tissues at those locations, and all possible regions of origin are sampled. An alternative approach is to characterize those predefined regions in terms of the distributions of isotope values taken from calibrated isoscapes. Once a calibrated assignment model is obtained, stable isotope measurements from individuals of unknown origin are used to determine the most likely region of origin. We describe three general forms of assignment methods: (1) classification trees, (2) likelihood-based methods with a uniform prior distribution, and (3) likelihood methods with structured prior distributions, and we discuss the utility of extending these models into a stochastic framework.

1. Classification Trees

A classification tree is a derived hierarchy of decision rules for assigning novel data to one of two or more classes. Each decision rule provides a fork or "branching event" in the flow toward assigning a data point to one of the predefined classes (regions). Classification trees do not require distributional assumptions and can combine discrete and continuous covariates. They rely on clustering algorithms or similar recursive computations. Thus, classification trees will be especially useful for future exploratory work and pilot studies, especially as more isotopes and trace elements are used because they accommodate not only multimodal distributions of isotopes but also work fine with any combination of continuous or discrete predictor variables.

Hebert and Wassenaar (2005a,b) employed this approach to assign mallards (*Anas platyrhynchos*) and northern pintails (*Anas acuta*) to one of four predefined geographic regions in North America using the values of δD, δ^{34}S, δ^{13}C, and δ^{15}N in feathers. Their application used univariate thresholds for decision branching. They tested the robustness of their model by assigning known-origin data not used to generate the classification tree. They did not, however, explore the robustness of the approach to how the potential geographic origins were predefined. As with all assignment methods discussed here, the efficacy of the approach will depend strongly on how the potential target geographic regions are defined.

The branching thresholds in classification trees can be fixed points along univariate gradients as in Hebert and Wassenaar (2005a,b) or they can be linear combinations of the predictor variables, as in discriminant function analysis. Rule sets derived from classification trees provide no specific information about how close the call is for a given split. In other words, it is very difficult to quantify the relative strength favoring one branch versus another. There are a host of software programs that employ various algorithms to ensure optimization of the trade-off between number of branching splits and predictive accuracy, but optimizations are just that—optimal solutions given the information at hand. If the likelihood for one region is nearly identical to that for another, there will still always be an optimal solution for differentiating between them, but this is not necessarily saying we have a great deal of confidence in doing so. As with any statistical modeling approach, this is just another way of saying that the performance of classification trees is limited by the quality and quantity of data.

2. Likelihood-Based Methods

One form of likelihood-based methods is discriminant analysis that can be used to classify a sample into one of two or more classes (regions). It has been used by several isotope-based studies to estimate the provenance of animals (Caccamise *et al.* 2000, Wassenaar and Hobson 2000, Farmer *et al.* 2004, Kelly *et al.* 2005, Rocque *et al.* 2006, Szymanski *et al.* 2007). Parameters for region-specific likelihood functions (always some form of the normal distribution) are then estimated from isotope data collected in each predefined region. The likelihood functions for each region are then evaluated for the isotope value measured for an individual of unknown origin, and the individual is then simply assigned to the region with the highest-valued likelihood (or probability). One advantage of likelihood-based methods is that they consider not only the mean isotope value of each region but also some measure of the variability.

Likelihoods can easily be converted to probabilities using *Bayes' Rule* (Table 5.1), which is a formal way to invert conditional probabilities. Bayes' Rule is simply an algebraic manipulation of an equality that springs from the definition of conditional probability. The probability of two events both happening, $P(A \text{ and } B)$ is the same as the probability of one event happening $P(A)$, joined with the probability of the second event happening, given that the first event occurred, $P(B|A)$. The vertical line means "given," or "conditioned on." In other words, $P(A \text{ and } B) = P(A) * P(B|A)$.

Because it does not matter which event occurs first, we can also write $P(A \text{ and } B) = P(B) * P(A|B)$. Recognizing that these two expressions are equal, we can write $P(B) * P(A|B) = P(A) * P(B|A)$, rearrange and get $P(B|A) = P(A|B) * P(B)/P(A)$, which is Bayes' Rule. In this expression, $P(B|A)$ is often referred to as the posterior probability of B given A, or the probability of some model parameters or hypotheses, given data. $P(A|B)$ is the likelihood function, or the probability of the observed data given some model parameters or hypotheses. $P(B)$ is referred to as the prior or marginal probability for B, or the probability of some model parameter without knowledge from data, and $P(A)$ is also a marginal probability that serves as a normalizing constant. $P(A)$ captures all possible outcomes for A, regardless of B. Because we have A conditioned on B, we sum or integrate over $P(A|B) * P(B)$. That is, we integrate over all possible outcomes where A occurs.

The use of probabilities using Bayes' Rule makes the interpretation of results much more straightforward than with likelihoods alone because we can talk directly about the probability of model parameters or hypotheses, rather than indirectly discussing the probability of observing the same data in a hypothetical replicate sampling event. In fact, discriminant analysis can be thought of as a special case of Bayesian analysis where the sampling probability distribution (the likelihood function) is normal, and the prior probability distribution is uniform over the candidate regions. Discriminant analysis is, therefore, naturally extended into a more Bayesian approach by giving structure to the prior probability distribution.

As an example, Royle and Rubenstein (2004) assumed normal distributions for the stable isotope values in feathers collected from each of three broad regions and argue that relative abundance is a good proxy for the prior probability that an individual of unknown origin came from any of those three regions. In other words, they point out that in the absence of any isotopic information, the probability of an individual originating from any given region is proportional to the relative abundance of animals from that region. Norris *et al.* (2006) also used relative abundance for the prior probability distribution, but they differed from Royle and Rubenstein (2004) by using precipitation-based isoscape values to define the sampling probability distributions for each region. In cases where relative population abundance estimates are not available, the number of individuals sampled from each potential area of origin can be used as a proxy for the prior probability (Wunder *et al.* 2005). In Table 5.3, we have outlined information sources in addition to relative abundance that could be integrated with isotopic data as prior probabilities in assignment tests. However, we caution against relying too strongly on such distributions unless researchers are confident that their sampling effort was even across all regions of potential origin; no study has thoroughly investigated the robustness of these likelihood-based methods to the way in which target regions are geographically defined.

Assuming normal distributions for the likelihood function is by no means a requirement. Given the relatively sparse nature of the data in these studies (as with many migratory studies), the assumption of normality was used as a way to marginalize the influence of outliers in the data and to simplify computations. If more data were available, fully Bayesian methods could have been applied to fuller effect. Full Bayes circumvents the need for any strong distributional assumptions, and can therefore also simultaneously accommodate both discrete and continuously distributed covariates.

For example, consider a simple case where stable isotope data are expected to follow a multimodal distribution: a predefined location may include a mix of C3 and C4 plants (and therefore a mix of $\delta^{13}C$ values). Suppose that the population of interest included two groups—one that foraged predominantly in C3-based food webs and one that foraged mostly in C4-based food webs. The distribution of $\delta^{13}C$ for that population would look bimodal. Calibrating an assignment model using a normal distribution would result in low assignment probabilities for individuals from either of the two groups that are actually known to be from that location because the probability density function of $\delta^{13}C$ was described

TABLE 5.3 Types of nonisotopic information that can be used as priori probabilities for assigning animals of unknown origin

Type of information	Source of information	Relevance to assigning birds using Baysian inference
Relative abundance	Large-scale surveys such as the Breeding Bird Survey or field data	Used because an individual of unknown origin already has a probability of originating from a given region based on relative abundance
Return rates	Field data on frequency of return in a population from one year to next	Could be used when estimating the natal origin of juveniles; probability of returning to natal grounds may also be related to distance
Mark-recaptures	Field data and large-scale coordinated efforts (*e.g.*, Bird Banding Laboratory)	Could be used if there is prior information on patterns of connectivity based on band recaptures; may not be useful if there are only a few recaptures
Migration distance	Previous studies on survival in relation to migration distance	Could be used as prior information if survival rates are known to vary with distance from potential regions of origin
Distribution range	Natural history	Constrains possible geographic origins to regions within known range
Phenotype	Museum skins, field data	Individuals with different phenotypes may be known beforehand to originate from specific regions; one caveat is that this only provides information on possible natal origins, does not accommodate possibility that an individual may have dispersed to an alternate location later in life
Genetics	mtDNA, AFLP, microsatellite	Same as above (including caveat) but for genetic composition
Isotopes	Light or heavy isotopes that vary geographically	May be particularly useful for studies that are updating or improving previous work
Trace elements	Chemical analysis of organic tissues	Same as above

as a normal (unimodal) distribution. Fully Bayesian approaches or classification trees would perform better in such situations. Classification trees, however, remain less useful for directly determining the relative probability of taking one branch versus the other. In contrast, a fully Bayesian analysis provides direct probability estimates and therefore also provides insight about the relative strength of assigning an individual to one region versus another. However, none of these assignment methods provide much insight into the mechanistic underpinnings of the functional relationship between isotopes and geography.

Partly in response to this problem, we recently used a simulation framework to explore the reliability of asymptotic results from Bayesian assignments, given stochastic fluctuations associated with IRMS measurements and with the process of spatially interpolating sparse data on δD in precipitation (Wunder and Norris 2008). There, we reevaluated a study by Norris *et al.* (2006) that calculated the mean and standard deviation of expected δD values for five predefined breeding regions. The δD values were extracted from spatially interpolated precipitation maps (Bowen *et al.* 2005) and standard deviation values ranged from 5‰ to 13.6‰. In this study, we considered not only the variation in these expected values from the interpolated precipitation-based δD isoscape but also the uncertainty associated with measuring δD in feathers and the uncertainty associated with the spatial interpolation itself. The latter portion of uncertainty captures the variation associated with how far a geographic point is away from an actual data sampling station; the further away a point is, the higher variability there will be in interpolating the δD value. Regions with high variability in any of the distributions of predicted δD likely reflected higher amounts of topographic relief and temperature variation over its range. Regions with high spatial interpolation uncertainty were associated with similar physiographic features, and also reflected the relative distance from a true data station where δD in precipitation was directly measured. Thus, the overall confidence of assignments to these regions was much lower.

This stochastic modeling extension was useful for exploring the relative sensitivity of conclusions to the assumptions that isoscapes are perfectly predicted and that δD in feathers is perfectly measured. It represents a preliminary, if not also rudimentary, approach toward studying the state of our understanding of the relationship between δD in feathers and geography. This method can be extended to include other factors known to influence the certainty of our modeling efforts. For example, we might consider the effects of reasonable stochastic fluctuations in the functional relationship between δD in precipitation and that in feathers (variable isotopic discrimination).

D. Probability Surfaces

Probability surfaces describe the probability that any point in space is the true origin of an individual for which stable isotope values are measured. In this way, probability surfaces provide a continuous response variable (typically represented as pixels/raster format on a map) constrained to values between zero and one. Because these are spatially continuous models, they rely on model output from spatially interpreted data. The nature of that model output determines the type of calibration that must be performed before proceeding to transform them into a spatially indexed probability distribution function. For example, if feather isotopes are to be compared against a precipitation-based isoscape, that isoscape must first be calibrated to reflect the isotope discrimination between feathers and precipitation. Alternatively, if the baseline process is an isoscape derived directly from the δ-values in feathers collected from across the spatial range of interest, then no further calibration is required—the calibration in such a case is the spatial interpolation itself.

The first crude effort to generate probability surfaces was recently attempted for determining the breeding origins of wintering mountain plovers (Wunder 2007). The basic algorithm used was to first calibrate the δD isoscape from Bowen *et al.* (2005) with feathers of known origin separately for both adult recaptures and fledglings. Next, a stochastic model was derived from three dependent (nested) variance-generating processes: (1) analytical error, estimated as described by Wunder and Norris (2008), (2) within-individual variance, differences within and among feathers estimated from values documented by Wassenaar and Hobson (2006); and (3) within-location variance, estimated from 112 values from the published literature and data collected for mountain plovers. These variance-generating processes are nested, in that a typical sampling process continues as follows: individual birds are sampled from a given location, which can be isotopically characterized by a probability distribution. That is, we first randomly sample individuals from the same location, each with some alternate mean δ-value. Then, we randomly sample a feather from each individual and we can then model among-feather variance. Finally, we measure the δ-value for the sampled feather, and we know this measurement is also estimated with error. The posterior distribution associated with the combination of these processes can be estimated using *Monte Carlo integration* (Table 5.1). Once this hierarchical model is parameterized, δD in feathers that were collected from birds on the wintering grounds (but grown during the breeding season) were used to evaluate the probability density function for each grid point in the calibrated isoscape. The result was a set of geographic maps showing the spatial distribution of the gradient in probability of origin for each individual.

In contrast to the binary response maps (ones and zeros) produced using map lookups by Lott and Smith (2006) and Hobson *et al.* (2007), probability surfaces are spatially explicit maps with values spanning the gradient from zero to one. Thus, they disclose the full range of possibilities based on estimable sources of error. For example, directly comparing the relative probability of any two points can be simply expressed in terms of odds ratios. This allows a great deal of flexibility for approaching problems from the full range of risk levels. There is no need to arbitrarily set threshold levels for the probability of making type I or type II errors. In probability surfaces, the probability space is geographically indexed, and not constrained to sample data as it is for the map lookup approaches discussed in Section VI.A that used tolerance limits.

This modeling approach is also useful in that it makes all sources of variance very transparent, providing insight into the most useful places for strategic experimentation. It is a simple matter via simulation to explore such questions about the relative gains for increased precision in any of the modeled variance-generating processes by iterating through the full parameter space for each process, while holding the others constant. This modeling approach holds great potential for providing back-and-forth dialogue between experimental and applied researchers that is necessary to advance isotope-based methods for determining geographic origin.

This use of probability surfaces is still relatively new and there is much room for further refinement. The approach is accessible to a wide range of study problems, but is currently most effectively done for migratory birds. Estimates of each source of variation need not be obtained for each specific study. Stochastic characterizations can be based on values in the literature. For example, Wunder (2007) used 112 estimates of within-population standard deviation, mostly from the literature, but only had 6 estimates for modeling the within-individual standard deviation, all from the literature on migratory birds. The primary disadvantage is that the procedure is relatively more computational intensive than either the map lookup method or likelihood-based assignments, but we believe that this approach provides the most robust models for estimating the origin of migratory animals over a continuous space.

VII. SAMPLING CONSIDERATIONS

As emphasized in this chapter, the most robust study designs will carefully consider the general idea of calibration, from IRMS calibration all the way up to assign-time calibrations of modeled spatial patterns at the continental or global scales. In the case of discrete assignment methods, each potential site or region of origin should ideally be isotopically calibrated with tissue known to have derived from each region. In the case of a more continuous analytical approach based on gradients, the gradients need to be calibrated using tissues of known origin from across the full extent of the gradient range. In other words, because of a relative lack of mechanistic understanding about when and where to expect strong deviations from otherwise smoothed patterns, the most robust approach is to determine these patterns empirically.

Equally important in all empirical characterizations of the relationship between stable isotopes and geography is the specification of all known sources of variance. We have seen that measurement error, within-individual variance, within-location variance, among-location variance, among-age class variance, and among-species variances all affect the relationship between isotopes and geography. Thus, sampling designs will do well to consider designs meant to isolate the effects of these factors.

VIII. SPATIAL CONSIDERATIONS

Calibration samples should ideally span the entire targeted geographic range of interest. That is, if the study seeks to infer the wintering grounds for a population of breeding birds, then the researchers should be sure to sample newly developed tissue from birds captured at the full suite of potential wintering locations across the wintering range. Alternatively, if using an isoscape, at a minimum, the extreme ends of the gradient for the isotope of interest need to be sampled. That is, suppose that the range of predicted δD values for the known wintering range of a species spans 100‰. In this case, at a minimum, the researchers would need to collect freshly developed tissue from the regions that represent the endpoints of that 100‰ range in order to develop a two-point (linear) calibration. Of course, a more ideal design would sample from far more than two locations and would explore the prospects of both linear and nonlinear calibrations.

IX. TEMPORAL CONSIDERATIONS

Distributions of stable isotope values for any given location may vary over time (over the course of a season or over multiple years). This aspect is less studied (Farmer *et al.* 2002), but the general hypotheses for this variation are quite logical. Some seasons and years provide more or less available energy than others. It is this flux in available energy that yields temporal differences in the discriminatory incorporation of heavy isotopes into various compounds and tissues. For many of the tissues of interest in the application of isotope ecology to migration, there is a relatively wide temporal window over which fluctuation is integrated. For example, it takes from 10 to 30 days to synthesize most avian feathers. Or waterfowl may be breeding in wetlands that dry up over the summer.

X. RELATIONSHIP BETWEEN SAMPLING AND PREDICTIVE RESOLUTION

This is an issue of error propagation and model choice. Predictive resolution and the necessary sampling resolution depend critically on model choice. Predictive resolution for models that use calibrated isoscapes or other interpolated surfaces is defined by the resolution of the spatial interpolation. Models in the family of discrete-response assignment methods allow the researcher to determine the predictive resolution to some degree.

For hydrogen isotopes, the best resolution that can be expected from any model using the δD isoscape of Bowen *et al.* (2005) is 0.33 degrees (www.waterisotopes.org). The δD isoscape of Meehan *et al.* (2004) (http://entomology.wisc.edu/~tdmeehan/ddp.html) is a δD grid with a 1-km resolution. Both model surfaces were generated from the same isotope training dataset for δD in precipitation (as provided by Global Network for Isotopes in Precipitation). The differences and subsequent relative utility of each of these isoscapes relative to various case studies have not yet been explored. Potentially, the δD isoscape of Meehan *et al.* (2004) provides a model of the pattern in particular (at fine local scales), but not in general, and Bowen *et al.* (2005) provides the opposite. If this is the case, then the Meehan isoscape will require localized calibration for nearly every study, whereas the Bowen isoscape may not.

With all discrete assignment models, the researcher predefines the potential target regions, and therefore also the predictive resolution of origin *a priori*. However, there are as yet undefined thresholds of resolution beyond which we will not be able to isotopically differentiate among regions. Although it is possible to increase the resolution at which regions can be differentiated by increasing the number of samples and the number of markers (isotopes, trace elements, etc.), such an approach will always require exhaustive sampling. This so-called shotgun approach may work for one set of sample data, but when it fails to generate similar results for another independent data set, the researcher is left wondering why. Experiments that effectively isolate mechanisms responsible for generating variance are especially useful for refining models that relate isotopes in tissue to geography.

XI. RELATIVE POPULATION ABUNDANCE CONSIDERATIONS

One of the primary goals in assigning migratory animals to geographic location using stable isotopes is to estimate overall patterns of connectivity between two or more periods of their annual cycle. Understanding these overall patterns will be important for predicting changes in population size and developing optimal conservation plans (Martin *et al.* 2007, Chapter 1). Here, estimates of relative population abundance during all periods of the annual cycle will play a critical role in our ability to link these patterns with predictive population models.

Consider a situation in which individuals are sampled for stable isotopes on the wintering grounds to infer their breeding origins. Twenty individuals are sampled at each of 15 sites throughout the wintering range. Assignment tests are then conducted on each individual and summarized for each site (see Section V above) to infer overall patterns of connectivity among wintering sites and breeding locations. However, it is more difficult to gain information about relative abundance if the sample size is fixed, rather than the sample effort. By fixing sampling effort (*e.g.*, same number of mist net hours at each site), it is easier to justify the assumption that sample size reflects relative abundance. If wintering sites are not sampled in proportion to their relative abundance, then inference from the modeled connectivity patterns may be somewhat misleading because it will overrepresent some sites and underrepresent others. Thus, one must be cautious when using these assignments to model the effects of habitat loss (Wunder and Norris 2008) or to determine how conservation plans in one season will influence population size the following season (Martin *et al.* 2007).

Information on relative population abundance from the period where animals are being assigned is also important. First, it can provide an indication of whether we have obtained a representative sample of individuals from the previous season. Following our example above, if we had perfect sampling coverage of the wintering grounds, then the proportion of birds assigned to each breeding region should match the relative abundance. If it does not then we can conclude one of two things: (1) we may have missed important areas in which the animals reside during the winter or (2) our sampling scheme may not have reflected the relative abundance on the wintering grounds. Second, as we have mentioned above (Section VI.C; Table 5.3), relative abundance can be used as a prior probability of origin in likelihood-based assignment tests (Royle and Rubenstein 2004). As long as prior probabilities of relative abundance do not dominate the generation of posterior probabilities (*i.e.*, when there are few differences in the isotopic data between potential regions of origin), then relative abundance can compliment isotopic information in a useful way. For example, when assigning animals to predefined regions, there can be individuals whose isotopic values effectively "ride the fence" between two or more regions. In this case, the incorporation of relative abundance acts as the tie breaker. It will assign that individual into whichever of the two regions has the highest estimated abundance.

XII. CONCLUSION

Stable isotopes offer a powerful new tool for tracking the movement of migratory animals. The landmark pioneering studies that used stable isotopes to track patterns of migration primarily relied on simple calibration rules and the map lookup approach (Chamberlain *et al.* 1997, Hobson and Wassenaar 1997). Over a decade later, major advances have been made in terms of both calibration and assignment models. In this chapter, we have provided a comprehensive overview of the methods that have been used to estimate the origin of migratory animals using stable isotopes. In doing so, we have attempted to expose the major assumptions associated with each model, show how sources of error can be better and more directly incorporated into these models, and emphasize the importance of understanding the mechanisms that drive these sources of isotopic variation. Our hope is that future research will adopt and further refine these approaches to derive robust models that can be successfully used to track migratory animals throughout their annual cycle.

XIII. REFERENCES

Bearhop, S., W. Fiedler, R. W. Furness, S. C. Votier, S. Waldron, J. Newton, G. J. Bowen, P. Berthold, and K. Farnsworth. 2005. Assortative mating as a mechanism for rapid evolution of a migratory divide. *Science* **310**:502–504.

Boulet, M. B., H. L. Gibbs, and K. A. Hobson. 2006. Integrated analysis of genetic, stable isotope and banding data reveal migratory connectivity and flyways in the northern yellow warbler (*Dendroica petechia; Aestiva* group). *Ornithological Monographs* **61**:29–78.

Bowen, G. J., L. I. Wassenaar, and K. A. Hobson. 2005. Global application of stable hydrogen and oxygen isotopes to wildlife forensics. *Oecologia* **143**:337–348.

Caccamise, D. F., L. M. Reed, P. M. Castellie, S. Wainwright, and T. C. Nichols. 2000. Distinguishing migratory and resident Canada Geese using stable isotope analysis. *Journal of Wildlife Management* **64**:1084–1091.

Chamberlain, C. P., J. D. Blum, R. T. Holmes, X. Feng, T. W. Sherry, and G. R. Graves. 1997. The use of isotope tracers for identifying populations of migratory birds. *Oecologia* **109**:132–141.

Dansgaard, W. 1964. Stable isotopes in precipitation. *Tellus* **16**:436–468.

Duxbury, J. M., G. L. Holroyd, and K. Muehlenbachs. 2003. Changes in hydrogen isotope ratios in sequential plumage stages: An implication for the creation of isotope base-maps for tracking migratory birds. *Isotopes in Environmental and Health Studies* **39**:179–189.

Farmer, A., R. Rye, G. Bern, C. Landis, C. Ridley, and I. Kester. 2002. Tracing the pathways of neotropical migratory shorebirds using stable isotopes: A pilot study. *Isotopes in Environmental and Health Studies* **39**:1–9.

Farmer, A., M. Fernandez, M. Abril, J. Torres, C. Kester, and C. Bern. 2004. Using stable isotopes to associate migratory shorebirds with their wintering locations in Argentina. *Ornitologia Neotropical* **15**:377–384.

Gröning, M. 2004. International stable isotope reference materials. Pages 874–906 *in* P. D. Groot (Ed.) *Handbook of Stable Isotope Analytical Techniques*, Volume 1., Elsevier, Amsterdam.

Hebert, C. E., and L. I. Wassenaar. 2005a. Feather stable isotopes in western North American waterfowl: Spatial patterns, underlying factors, and management applications. *Wildlife Society Bulletin* **33**:92–102.

Hebert, C. E., and L. I. Wassenaar. 2005b. Stable isotopes provide evidence for poor pintail production on the Canadian prairies. *Journal of Wildlife Management* **69**:101–109.

Hobson, K. A. 2005. Stable isotopes and the determination of avian migratory connectivity and seasonal interactions. *Auk* **122**:1037–1048.

Hobson, K. A., and L. I. Wassenaar. 1997. Linking breeding and wintering grounds of neotropical migrant songbirds using stable hydrogen isotopic analysis of feathers. *Oecologia* **109**:142–148.

Hobson, K. A., L. I. Wassenaar, and O. R. Taylor. 1999. Stable isotopes (δD and δ^{13}C) are geographic indicators of natal origins of monarch butterflies in eastern North America. *Oecologia* **120**:397–404.

Hobson, K. A., K. P. McFarland, L. I. Wassenaar, C. C. Goetz, and J. E. Rimmer. 2001. Linking breeding and wintering grounds of Bicknell's thrushes using stable isotope analysis of feathers. *Auk* **118**:16–23.

Hobson, K. A., S. V. Wilgenburg, L. I. Wassenaar, H. Hands, W. P. O'Meilia, M. Johnson, and P. Taylor. 2006. Using stable hydrogen isotope analysis of feathers to delineate origins of harvested sandhill cranes in the central flyway of North America. *Waterbirds* **29**:137–147.

Hobson, K. A., L. I. Wassenaar, F. Moore, J. Farrington, and S. V. Wilgenburg. 2007. Estimating origins of three species of neotropical migrant songbirds at a gulf coast stopover site: Combining stable isotope and GIS tools. *Condor* **109**:256–267.

Jardine, T. J., and R. A. Cunjak. 2005. Analytical error in stable isotope ecology. *Oecologia* **144**:528–533.

Kelly, J. F., V. Atudorei, Z. D. Sharp, and D. M. Finch. 2002. Insights into Wilson's warbler migration from analyses of hydrogen stable-isotope ratios. *Oecologia* **130**:216–221.

Kelly, J. F., K. C. Ruegg, and T. B. Smith. 2005. Combining isotopic and genetic markers to identify breeding origins of migrant birds. *Ecological Applications* **15**:1487–1494.

Langin, K. M., M. W. Reudink, P. P. Marra, D. R. Norris, T. K. Kyser, and L. M. Ratcliffe. 2007. Hydrogen isotopic variation in migratory bird tissues of known origin: Implications for geographic assignment. *Oecologia* **152**:449–457.

Lott, C. A., and J. P. Smith. 2006. A geographic-information-system approach to estimating the origin of migratory raptors in North America using stable hydrogen isotope ratios in feathers. *Auk* **123**:822–835.

Lott, C. A., T. D. Meehan, and J. A. Heath. 2003. Estimating the latitudinal origins of migratory birds using hydrogen and sulfur stable isotopes in feathers: Influence of marine prey base. *Oecologia* **134**:505–510.

Martin, T. M., I. Chades, P. Arcese, P. P. Marra, H. P. Possingham, and D. R. Norris. 2007. Optimal conservation of migratory birds. *Public Library of Science, ONE* **2**(8):e571.

Mazerolle, D. F., and K. A. Hobson. 2005. Estimating origins of short-distance migrant songbirds in North America: Contrasting inferences from contrasting hydrogen isotope measurements of feathers, claws and blood. *Condor* **107**:280–288.

Mazerolle, D. F., and K. A. Hobson. 2007. Patterns of differential migration in white-throated sparrows evaluated with isotopic measurements of feathers. *Canadian Journal of Zoology* **85**:413–420.

Mazerolle, D. F., K. A. Hobson, and L. I. Wassenaar. 2005. Stable isotope and band-encounter analysis delineate migratory patterns and catchment areas of white-throated sparrows at a migration monitoring station. *Oecologia* **144**:541–549.

Meehan, T. D., C. A. Lott, Z. D. Sharp, R. B. Smith, R. N. Rosenfield, A. C. Stewart, and R. K. Murphy. 2001. Using hydrogen isotope geochemistry to estimate the natal latitudes of immature Cooper's Hawks migrating through the Florida Keys. *Condor* **103**:11–20.

Meehan, T. D., R. N. Rosenfield, V. N. Atudorei, J. Bielefeldt, L. J. Rosenfield, A. C. Stewart, W. E. Stout, and M. A. Bozek. 2003. Variation in hydrogen stable-isotope ratios between adult and nestling Cooper's hawks. *Condor* **105**:567–572.

Meehan, T. D., J. T. Giermakowski, and P. M. Cryan. 2004. GIS-based model of stable hydrogen isotope ratios in North American precipitation for use in animal movement studies. *Isotopes Environ Health Study* **40**:291–300.

Norris, D. R., P. P. Marra, R. Montgomerie, T. K. Kyser, and L. M. Ratcliffe. 2004. Reproductive effort, molting latitude, and feather color in a migratory songbird. *Science* **306**:2249–2250.

Norris, D. R., P. P. Marra, G. J. Bowen, L. M. Ratcliffe, J. A. Royle, and T. K. Kyser. 2006. Migratory connectivity of a widely distributed songbird, the American redstart (*Setophaga ruticilla*). *Ornithological Monographs* **61**:14–28.

Paxton, K. L., C. V. Riper III, T. C. Theimer, and E. H. Paxton. 2007. Spatial and temporal migration patterns in Wilson's warbler (*Wilsonia pusilla*) in the southwest as revealed by stable isotopes. *Auk* **124**:162–175.

Perez, G. E., and K. A. Hobson. 2006. Isotopic evidence of interrupted molt in northern breeding populations of the loggerhead shrike. *Condor* **108**:877–886.

Rocque, D. A., M. Ben-David, R. P. Barry, and K. Winker. 2006. Assigning birds to wintering and breeding grounds using stable isotopes: Lessons from two feather generations among three intercontinental migrants. *Journal of Ornithology* **147**:395–404.

Royle, J. A., and D. R. Rubenstein. 2004. The role of species abundance in determining breeding origins of migratory birds with stable isotopes. *Ecological Applications* **14:**1780–1788.

Rubenstein, D. R., C. P. Chamberlain, R. T. Holmes, M. P. Ayres, J. R. Waldbauer, G. R. Graves, and N. C. Tuross. 2002. Linking breeding and wintering ranges of a migratory songbird using stable isotopes. *Science* **295:**1062–1065.

Smith, R. B., T. D. Wolf, and B. O. Meehan. 2003. Assessing migration patterns of sharp-shinned hawks *Accipiter striatus* using stable-isotope and band encounter analysis. *Journal of Avian Biology* **34:**387–392.

Smith, A. D., and A. M. DuftyJr.. 2005. Variation in the stable-hydrogen isotope composition of northern goshawk feathers: Relevance to the study of migratory origins. *Condor* **107:**547–558.

Szymanski, M. L., A. D. Afton, and K. A. Hobson. 2007. Use of stable isotope methodology to determine natal origins of mallards at a fine scale within the upper Midwest. *Journal of Wildlife Management* **71:**1317–1324.

Wassenaar, L. I., and K. A. Hobson. 2000. Stable carbon and hydrogen isotope ratios reveal breeding origins of red-winged blackbirds. *Ecological Applications* **10:**911–916.

Wassenaar, L. I., and K. A. Hobson. 2001. A stable-isotope approach to delineate geographical catchment areas of avian migration monitoring stations in North America. *Environmental Science and Technology* **35:**1845–1850.

Wassenaar, L. I., and K. A. Hobson. 2002. Comparative equilibrium and online technique for determination of non-exchangeable hydrogen for keratins for use in animal migration studies. *Isotopes in Environmental and Health Studies* **39:**211–217.

Wassenaar, L. I., and K. A. Hobson. 2006. Stable-hydrogen isotope heterogeneity in keratinous materials: Mass spectrometry and migratory wildlife tissue subsampling strategies. *Rapid Communications in Mass Spectrometry* **20:**2505–2510.

Wunder, M. B. 2007. Geographic Structure and Dynamics in Mountain Plover. Ph.D. Dissertation. Colorado State University, Fort Collins.

Wunder, M. B., and D. R. Norris. 2008. Improved estimates of certainty in stable isotope-based methods for tracking migratory animals. *Ecological Applications*, in press.

Wunder, M. B., C. L. Kester, F. L. Knopf, and R. O. Rye. 2005. A test of geographic assignment using isotope tracers in feathers of known origin. *Oecologia* **144:**607–617.

CHAPTER 6

Future Directions and Challenges for Using Stable Isotopes in Advancing Terrestrial Animal Migration Research

Jeffrey F. Kelly,* Stuart Bearhop,[†] Gabriel J. Bowen,[‡] Keith A. Hobson,[§]
D. Ryan Norris,[¶] Leonard I. Wassenaar,[§] Jason B. West,** and Michael B. Wunder[††]

*Oklahoma Biological Survey and Department of Zoology, University of Oklahoma Norman
[†]Center for Ecology and Conservation, University of Exeter
[‡]Earth and Atmospheric Sciences Department, Purdue University
[§]Environment Canada
[¶]Department of Integrative Biology, University of Guelph
**Department of Biology, University of Utah
[††]Department of Fish, Wildlife, and Conservation Biology, Colorado State University

Contents

I. INTRODUCTION

It is clear that a detailed understanding of animal movements underpins a host of fundamental research topics, ranging from the evolution of life histories, behavioral plasticity, and constraints to the genetic basis for migration and dispersal (Chapter 1). However, it has been pointed out on numerous occasions that our great lack of information on dispersal and migration represents one of the most formidable

Tracking Animal Migration with Stable Isotopes
K. A. Hobson and L. I. Wassenaar (Editors)
ISSN 1936-7961, DOI: 10.1016/S1936-7961(07)00006-1

gaps in our knowledge. This gap sits at the core of our understanding of animal ecology and evolutionary biology (*e.g.*, Koenig *et al.* 1996). Surprisingly, the obstacle for closing this information gap is primarily a technological one rather than a conceptual one (Wikelski *et al.* 2007). Breakthroughs derived from technological advances will no doubt also shed new insight into theoretical and conceptual models at both the individual, and population level. Simply put, technological advances in animal tracking has and will continue to vastly improve our understanding of nearly every aspect of organismal, population, and community ecology.

The advent of stable isotope methods as well as other intrinsic markers has coincided with important developments in the technology of animal tracking using extrinsic markers. Thus, we find ourselves at a point in time where new and diverse developments are occurring on a number of research fronts. In this chapter, we examine the path ahead and, in particular, how stable isotope approaches can be best combined with other promising and new techniques.

II. THEORETICAL CONTEXT

While we have argued in this volume for the general case that stable isotope approaches will continue to bear fruit in studying animal migration, we believe that even more progress can be made by research that is very specific in its objectives. That is, we should focus on key questions that have the highest priority for advancing our knowledge of dispersal and migration ecology. Keeping a clear focus on key questions will help avoid the pitfalls of following methodological and shotgun approaches that may provide little insight into questions of biological significance.

Recently, deliberations from a network of migration biologists have proposed four key research questions (see Migrate website at http://www.migrate.ou.edu) that we also think require immediate attention. These four research questions are given below:

1. What are the determinants of behavioral plasticity in migrant animals and what are the constraints on behavioral adaptation? For example, how do individual migrants cope with, say, weather-induced uncertainty over the short-term, and under what circumstances are populations not able to adapt to long-term environmental changes?
2. What are the determinants of individual fitness of migrants? In particular, knowing when and where most mortality occurs and the environmental and biotic events that contribute most to variation in mortality rates would be very useful.
3. What are the drivers of population dynamics in migratory animals? Having good estimates of vital rates that could be used for population projections and understanding which vital rates are most sensitive (or elastic) would be a major advance.
4. What is the impact of environmental change on migratory life histories? In particular, what are the impacts of land use, environmental, and climatic changes? Ultimately, critical evaluation of predictions regarding changes in population dynamics and individual behavior and adaptation in response to land use and the environment will be the measure of advance in our science.

The spatial and temporal extent of research studies designed to examine the role of behavioral plasticity will likely differ from those that examine long-term impacts of environmental change. Nonetheless, it is useful to have these questions clearly in mind during discussions of how to improve (1) the kinds of isotope data we collect and (2) the geospatial information that we can extract from those data. All of the above questions require knowledge of how individuals and populations are spatially connected between all periods of their annual cycle. Without knowledge of where individuals are spending different periods of the year, it is impossible to infer the factors that influence behavioral plasticity, what are the primary factors that influence fitness, how populations are limited and regulated, and how environmental change can affect life-history tactics. In this context, we examine some outstanding challenges and limitations that are needed to address these evolutionary, ecological, and conservation questions in the future. We place special emphasis on those areas where innovation is needed to overcome known obstacles.

III. CHALLENGES AND LIMITATIONS

The following discussion is divided into two interrelated questions that portray the challenges and limitations of trying to predict the origins of animals from stable isotope values obtained from their tissues. First, how can we improve the collection and analysis of stable isotope data to infer origins of a particular migrant or population of migrants? Second, how can we extract more information from the isotope data we collect so that we optimize our ability to assign migrants to sites of origin? These two issues are at the heart of understanding the ways in which we can increase the utility of isotopes as spatial markers and what obstacles might lay in the way.

We know that significant sources of variation exist when using stable isotopes to track migratory individuals. Asking what would be an "ideal situation" can provide some guidance as to what components of variation we should be addressing. An ideal situation for tracking migratory animals would be one where we were able to sample all of the potential sites of migrant origin, all isotopic data had very high degree of spatial structure, no within-site variation, and every individual had the same diet and physiology (and thus the same diet-tissue discrimination factor). Ideally, the tissue sampled 100% reliably grown at a known location and would be biologically inert once grown. The ideal situation would also have no seasonal variation in the isotopic values of the energy and hydrological cycles that fuel primary producers and drive the food web. And, instrumental error would be zero.

Obviously, fully ideal situations do not exist but we can deal with these factors causing variation in two ways. We need (1) a better understanding of the mechanisms that drive the observed isotopic variation so we can adjust the design of future studies and (2) to incorporate this variation into our statistical methods for assigning individuals to places of origin. We address both of these topics below.

A. Toward Better Isotope Data

What are "better" isotope data? To us, this means isotope data from a migrant animal's tissues that provides the most unambiguous indicator of the geographic location at which the tissues were grown. As discussed in Chapter 3, it will be crucial to choose the appropriate tissue that reflects the appropriate period of dietary integration of interest. We also need to know what ecological and physiological factors determine the diet-tissue isotopic discrimination factor corresponding to that tissue and how that discrimination varies both within and among individuals at a given location or on a given diet. While several studies have provided useful working estimates for a number of tissues representing a number of species, especially for $\delta^{15}N$ and $\delta^{13}C$, we now know that diet quality and physiological state of an animal can influence such factors. It is also clear that we have very little idea of how migration per se can affect the behavior of isotopic signals in animals because few studies have been able to mimic migration as part of the experimental protocol (Hobson and Yohannes 2007). This is true also for estimates of isotopic turnover rates in tissues of exercising animals appropriate to applications involving individuals in the wild. The situation becomes even more complex when we consider the use of hydrogen isotope measurements in tissues because of the additional opportunity for hydrogen isotope exchange within animals and for the distinct roles of diet and drinking water that will influence tissue isotope measurements.

There has been a long-standing call for improved and experimentally based foundations for stable isotope approaches (*e.g.*, Gannes *et al.* 1997). We must consider more directly what is required to improve isotopic studies of migration. Captive rearing experiments are needed that will pay far more attention to the composition of diets with a view to replicating as closely as possible the nutritional range experienced by migrating animals of interest (*e.g.*, Pearson *et al.* 2003). Researchers using $\delta^{15}N$ and $\delta^{13}C$ must consider and report the elemental C:N ratios for the foods they use because this will give an indication of the protein content and assist with estimates of the appropriate nitrogen isotope discrimination factor to use. Recent research has also demonstrated the value of considering metabolic pathways associated with the macromolecules of proteins, lipids, and carbohydrates over bulk tissues. Here, it may be most useful to consider all possible sources of an element in the tissue of interest. We also require more information on how prolonged exercise associated with migration affects isotopic discrimination and elemental turnover rates, studies best accomplished using wind tunnel experiments for example. Similarly, effects of age and sex, especially related to physiological condition or reproduction, would be useful areas of research using stable isotopes.

These developments underscore the need to better combine disciplinary expertise in animal physiology and biochemistry together with that of ecology and stable isotopes. The more closely we can mimic physiological conditions related to migration, reproduction, molt, and migratory preparation in captive animals along with predictive isotopic models, the better we will be able to model behavior of stable isotopes in wild migratory animals. This rigor is especially needed for δD and $\delta^{18}O$ for which little is known about physiological and dietary factors determining the isotopic composition of animal tissues. In the short term, emphasis should be placed on fixed keratinous tissues and blood components because these two are proving to be of the greatest use in migration tracking. However, understanding factors that determine δD and $\delta^{18}O$ values of muscle, lipid, liver, and bone collagen will be of importance. So, we can look forward to the development of a series of models that carefully link tissue-specific δD and $\delta^{18}O$ values of organisms with predicted or measured surface water, ground water, or mean precipitation values. There are many emerging examples in which physiology, apart from habitat choice and diet, is crucial to understanding the isotope ratios of an individual's tissues (Chapter 3). Models that couple physiological processes of consumers with the physical processes that drive patterns in stable isotope ratios in the environment will be a key for improving the precision of geospatial assignments. These physiologically informed models can be directly incorporated into probability density functions that can be used to map the origin of migrants. Thus, we can imagine models that take the available geospatial isoscape data and combine it with information on body size, tissue type, level of exercise, species group, and perhaps migratory diet to assist us in making the most of our isotope data.

Once we have increased our knowledge about the fundamental currencies of diet-tissue isotopic discrimination and elemental turnover for organisms of interest, we can consider other aspects that include refining our field methods and experimental designs. So far, good progress has been made on providing estimates of these parameters for birds, but far more work is required to explore sources of variation. Very little data exist for other animals such as insects and mammals.

B. Problems of Confidence and Precision in Geospatial Assignment

As developed by Wunder and Norris (Chapter 5), the ideal situation to infer origins of migratory animals from isotopic analyses of their tissues will be to have as complete a picture of the isoscape from which that animal's tissues were derived. This picture includes a good appreciation of the degree and causes of isotopic variation at single locations. As a minimum, studies should attempt to estimate the isotopic variance among individuals from known locations for each tissue of interest, within tissue variance for the species of interest, laboratory measurement error, and an estimate of the error associated with the isoscape model being used. The use of nonisotope information about the relative probability of origin is also encouraged. A Bayesian framework provides a way to formally quantify multiple sources of information about the relative probabilities of origin, but insights can also be gained from stepwise analysis that iteratively partitions the geospatial extent of potential origins.

Ongoing developments in GIS applications and remote sensing information layers will be extremely useful for narrowing possible sites of origin. An obvious example would be a species that only occupied forested habitat or those only occurring above a certain elevation. As habitat requirements become better understood for various taxa, those habitat types can be mapped using spatially explicit datasets. Narrowing possible areas of origin from more detailed information will help researchers to (1) develop sampling strategies for developing robust species-specific isoscapes and (2) reduce the potential amount of regional variation of isotopes in environmental samples.

Researchers should also consider optimal sample sizes of individuals that will provide the best estimate of different components of variance. In this sense, the isotope approach is no different from any other type of ecological sampling. Unlike most other ecological sampling, however, all such estimates are confounded by measurement error. That is, the δ-values are not known exactly, so estimates of population standard deviations based on the observed δ-values are always "underesti-mates." Other than Wunder (2007) and Wunder and Norris (2008), the impact of measurement error has not yet been fully explored. Field studies that document the isotopic variance in populations of organisms from single locations and that attempt to tease out the environmental or physiological causes of that variation are very much needed. In addition to captive studies, meta-analyses of datasets across species and isoscapes may prove to be useful in identifying and better understanding between-individual variation in tissue isotope values.

There are also the challenges of integrating spatial demographic data with isoscapes to arrive at a continuous picture of migratory connectivity. However, we are often more interested in migratory dispersal and connectivity *among* populations. That is, the rates at which demographic units exchange individuals (Salomonson 1955, Webster *et al.* 2002, Chapter 5). These rates can often be estimated with more certainty than can the geographic coordinates from which an individual originated. However, most previous studies of migratory connectivity have used arbitrarily defined regions or populations to develop assignment likelihoods (*e.g.*, Kelly *et al.* 2005). Few studies have attempted to add demographic parameters to their models. As we begin to try to merge information on precipitation isotope ratios, tissue isotope ratios, and organismal demographic data, concerns about matching the proper temporal scales are exacerbated. The more information available on the demographic structure of populations, the more we will be able to define appropriate population boundaries for use in assignment using stable isotope approaches. Such demographic data might be provided with the careful use of genetics (see below).

As modeling frameworks become more technologically sophisticated, the challenge of making them and the results accessible to a wide range of research and wildlife management audiences becomes important. Many of the user and client groups who would most benefit from applications of advanced modeling approaches have difficulty accessing (or even understanding) these new approaches. In order to expedite both the development and distribution of novel models, it will be important for modelers to work alongside ecologists, physiologists, and biochemists to push the frontiers of the field, and to clearly communicate the findings to environmental groups and managers.

C. Stable Isotope Measurements

There are a number of needs for improving the analytical methods used in stable isotope studies. Most of these relate to hydrogen isotope exchange issues in keratins and other organic tissues (Chapter 2). One pressing issue is the need for organic tissue laboratory standards where the hydrogen isotope ratios have been validated in multiple laboratories that apply similar sample preparation methods. Another problem is the lack of organic H standards spanning the positive and negative ends of range of values, encountered in migrant animals. The absence of widely available standards spanning the entire isotopic range contributes bias in the correction for exchangeable and nonechangeable hydrogen at both ends of the observed range. A third issue is the lipids within and external to the sampled tissues. It is clear that decisions on what solvents are used to clean tissues can have significant effects on the measured isotope values (Post *et al.* 2007). However, there is no consensus on the best methods for treatment of particular tissues prior to isotope analyses.

It is possible that other advances could be made through compound-specific isotope analysis (CSIA) of amino acids, fatty acids, or other compounds within specific tissues. Analysis of the isotope ratios of specific compounds have the potential advantage of providing a more direct linkage between the environment (*i.e.*, via diet) and the migrant's tissues. If the appropriate compound were chosen, the potential to improve the geographic signal from the environment may be considerable. For example, essential amino acids do not appear to undergo isotopic discrimination between diet and a consumer's tissue. Thus, in cases where the diet or trophic level of the organisms is unknown, the use of these essential amino acids could reduce variance in baseline food web signatures, especially for δ^{13}C and δ^{15}N measurements. Compound-specific analyses using deuterium measurements of animal tissues are in its infancy and it would be extremely useful if there were similar compounds involving hydrogen and oxygen that could provide information on baseline water isotope values without the complications of physiological reprocessing, isotope discrimination, and isotope exchange within the animal. To date, there have been relatively few compound-specific isotope studies with demonstrable links to migration research. However, Popp *et al.* (2007) recently demonstrated the utility of distinguishing between "trophic" amino acids (*i.e.*, those showing enrichment with trophic level such as alanine, aspartic acid, and glutamic acid) and "source" amino acids (glycine and phenylalanine) in delineating aspects of the ecology of migratory tuna (*Thunnus albacares*). Gleixner and Mugler (2007) recently demonstrated the potential in using compound-specific hydrogen isotope analysis of *n*-alkanes to infer past climate from lake sediments. The use of compound-specific hydrogen isotope analyses in migratory wildlife would represent a completely new area of research that may be able to explain some of the variance observed among individuals. Other areas of interest for researchers of animal movements will be the use of compound-specific measurements to identify point-source compounds or pollutants of anthropogenic origin (Evershed *et al.* 2007). As noted in Chapter 1, the use of spatial patterns in the occurrence or concentrations of contaminants and other compounds in the environment can provide information on origins of migratory animals. The addition of CSIA may well improve the resolution of such approaches, and researchers should always be cognizant of the delineation of spatial patterns in any environmental compound, should there be an appropriate application to tracking migrant animals.

There are caveats that go along with this call for new compound-specific assays. The isolation and analysis of specific compounds are comparatively difficult, and therefore far more expensive than bulk tissue isotope analyses. Also, CSIA are exceedingly time consuming and thus it may never be possible to achieve the rapid throughput or to facilitate the numbers of samples that allow us to generate the large sample sets needed for statistical analyses of large-scale sampling efforts. If the dietary sources of the specific compound, for example an essential amino acid, are not more homogeneous in their isotopic composition than the bulk diet, then no increased resolution would be expected. Finally, in the case of hydrogen, if the portion of the specific compound of interest (*e.g.*, the H atoms of an essential amino acid) is exchangeable with those in the environment, then there will be substantial variance introduced into the process. These caveats have been largely responsible for the slow development of CSIA despite several successful applications to ecological questions (Evershed *et al.* 2007).

D. Improving Isoscape Models

The utility of isoscapes rests on our understanding of the fractionating processes that occur between ingestion of the animal's water or food source and the production of the animal tissue or compound being measured, and the accuracy with which we can map the spatial variability of that source. Therefore, improvements in models describing and mapping this spatial variability, as well as models describing fractionations associated with animal metabolism will significantly improve the utility of isoscapes to address questions in animal ecology and evolution. While it would be ideal if all potential areas of origin differed categorically in their stable isotope values, this will almost never be the case across an entire range of a species, especially when attempting to differentiate between climatically or geologically similar regions. Despite this real-world limitation, there is significant potential for improvement in assignment of migratory individuals and populations using isotopes through increases in the accuracy of isoscape models and improved statistical assignment techniques (Chapter 4 and 5).

What are some ways that existing models can be improved to extract more information from the available data? One possibility is through downscaling of isoscape models to a resolution of kilometers, meters, and perhaps finer scales, although as noted in Chapters 4 and 5, this approach must be carefully balanced by concurrent increases in the resolution of the data and spatial specificity of isoscape models. There are multiple challenges here that relate to problems of small-scale heterogeneity and medium-scale homogeneity in isoscapes (Chapter 4 and 5). The stable isotope ratios of both N and C can vary over just a few meters, while H and O isotopes scale over large areas, even within biogeographic regions that share similar δD and $\delta^{18}O$ values. These scaling issues have an important impact on the certainty by which we can assign origins. Combining multiple isotope ratio measurements (*e.g.*, C, N, H, and O) may in some cases improve model predictions, especially when elements whose isotopic variations are discrete (*e.g.*, Sr) are combined with those with more continuous variation (*e.g.*, H, see Chapter 4). Work in the future should address these limitations directly, potentially improving the information that isoscapes and tissue- or compound-specific isotope ratio analysis yield.

An additional avenue that is just beginning to be explored is the improved characterization of isotopic variation in food webs over temporal scales. Thus, an additional unanswered question is how can the observed isotopic variation among and within years be used to improve predictions for migratory species over specific time frames? These temporal challenges will initially require specific assumptions about the mechanisms and time lags that influence isotope ratios of animal tissues, but should also spur important mechanistic research to improve our understanding of these dynamics. As this work progresses, there is significant potential for the development and application of improved isoscape models that capitalize on temporally resolved satellite and climate data to predict spatial isotopic patterns for discrete temporal intervals of relevance to specific migration research problems.

E. Linking Stable Isotopes and Other Geographic Markers

There are numerous intrinsic and extrinsic markers that can add some level of geographic information to stable isotope approaches. It follows that the certainty of assignment of a migrant's origin may (but not always!) be improved by linking isotopic patterns with geographic information from other sets of nonisotopic intrinsic or extrinsic markers. While the ability of other markers to provide geographic information is well established, there has been little rigorous work on defining the limits of resolution for each marker or assessing the possible benefits and accuracy of combining multiple (isotopic and non-isotopic) markers. Evaluating the limitations of each combination of intrinsic markers will be important for understanding which types of questions are best addressed with which combinations of markers.

The potential to combine innovations in intrinsic and extrinsic markers to track individuals and populations opens a wide array of exciting research questions that to date have been impossible to address. First, extrinsic markers could be combined with isotopes to improve the ability to assign individuals of unknown origin. One of the most obvious avenues that has not yet been explored in a rigorous manner is integrating information obtained from stable isotopes with band recovery data. For example, a few individuals of a species may be marked on the breeding grounds and recovered on the wintering grounds. This information could be used as a prior probability of origin before assigning wintering individuals to breeding areas using isotopes. In this case, information on band recoveries need not be confined to the wintering period. Recoveries during migration can provide useful prior information about the general direction of migration and, therefore, help resolve assignments where isotopes do not have spatial resolution to do so.

The potential to combine innovations in intrinsic and extrinsic markers to track individuals and populations opens a wide array of research questions that, to date, have been impossible to address. The rarity of such studies reflects the degree to which technological hurdles prevent most researchers from employing this strategy. Certainly, any study employing satellite or other tags that produce precise journeys or origins of individuals should, if the individual is recovered, sample those tissues for which the provenance of the period of integration is known. This would provide a means of determining how well our isoscape models relate to reality, a type of "field validation" or "ground truth" that has been impossible to obtain so far.

Genetic markers are inherited and fixed for the life of an individual, whereas isotopic markers are generated continuously. Thus, these two markers provide different (and potentially complementary) kinds of information. Genetic markers can provide information about the natal population for a migratory animal whereas isotopic markers provide only information about the recent origin of an individual from the previous season. This means that these two types of information could provide different estimates of origin. In such cases, the individuals that do not seem to "fit" could be very informative. For example, if the genetic data suggest membership in population A, but the isotope data suggest membership in population B, one possible explanation is that the individual was born into population A, but dispersed into population B. On the other hand, using genetics to estimate recent patterns of connectivity between seasons may be slightly misleading. Applying genetic markers for this purpose should ideally incorporate the probability that an individual greater than 2 years of age dispersed away from its natal site to a new breeding area. The other obvious limitation in using genetic markers is that individuals of unknown origin can only be assigned to the breeding period which means they are less useful when used to assign individuals to other periods of the annual cycle.

Present obstacles to better integration of isotopic and genetic data derive from three main sources. First, the number of genetic markers for which it is possible to examine geographic patterns is enormous. Much of the existing research has relied on mitochondrial DNA phylogeography. It seems likely that patterns detected with these markers could be improved through use of amplified fragment length polymorphism (Irwin *et al.* 2005) and single nucleotide polymorphism methods (Bensch *et al.* 2002). However, methodologies and the number of possible genetic markers continue to expand exponentially making it difficult to develop an efficient method of integrating genetic results with stable isotope ratios. A second problem is that most genetic markers can be measured more efficiently when the DNA samples are isolated from blood that creates some tissue mismatches for isotope analyses based on keratins. This tissue mismatch can potentially be useful in measuring dispersal. Finally, genetic data, by definition, is not continuous but rather categorical which requires different approaches for geospatial representation (Guillot *et al.* 2005, Manni *et al.* 2004). For this reason, the integration of genetic and isotope data in geospatial analyses has lagged development in these separate fields. Progress will require collaboration between isotopists and geneticists with expertise in geospatial modeling (a rare combination!). Any progress arising from this collaboration would have an immediate impact.

In particular, we appear to be at the beginning of a proliferation in the use of stable isotope methods to answer questions about dispersal ecology. The basic idea is that dispersing animals would appear as outliers (genetically, isotopically, or both), after considering the full range of variation estimated for a given location. One difficulty with this approach is that individuals sometimes grow feathers at alternate points in the life cycle. For example, we might assume a feather is grown during the late breeding season on the breeding grounds, when it was actually grown during migration. To separate these cases from true dispersal, having a unique population genetic marker would be useful. Immediate dispersers would have both rare isotope ratios and rare genetic haplotypes compared with the rest of the breeding population. In a similar way, if the isotope data suggest membership in a breeding population (A), but the genetic data suggest membership in a different population (B), this combination of results might suggest that dispersal from population B into population A occurred in a previous year. Demographic information on return rates or apparent survival is another type of information that can be readily incorporated with isotopic and genetic data to estimate breeding dispersal. For example, if we find that 60% of marked adults return to their breeding site between years, we can use this estimate as a prior probability before assigning individuals using intrinsic markers. Similarly, we may have prior information from marked individuals on dispersal distances that can be easily incorporated into probabilistic assignment tests.

Several previous papers have alluded to the use of trace element and heavy isotope analyses to infer origins of migratory animals. To date, trace element profiles, where a suite of elements are measured for relative abundance, have shown significant variation among subpopulations and so appear to have both potential and complications. This is especially the case for species that are known to congregate in a few breeding or wintering sites. Problems arise for cases where species are more diffusely spread across their ranges. Here we face the problem of having little *a priori* information on existing or expected elemental profiles corresponding to origins. Clearly, a much closer collaboration between researchers of animal migration and geologists or earth scientists familiar with regional or continental patterns of trace elements is required. Rather than considering a suite of elements in a multivariate sense, it is likely to be more productive on a case-by-case basis to consider the potential of geographical origins using a few key elements. We can imagine scenarios where deuterium analyses provide reasonable constraints on latitude of a migratory population and some trace elements (like copper or zinc) then provide additional information on origins within that range. The usual caveats of ground-truthing of elemental signatures of organisms from known locations still apply.

Our emphasis in this volume has been on the use of the light stable isotopes (C,N,H,O,S). A few studies have also used Sr isotope measurements because these vary geographically according to the age of the substrate and also differ between marine and terrestrial origins. The stable isotopes of Pb also show considerable potential to provide information on location because this element too is known to vary isotopically regionally. Whereas analyses of the heavier stable isotopes have been the domain of a few highly specialized laboratories, new technological developments (*e.g.*, inductively coupled plasma mass spectrometry) have now opened up the more common use of these and a number of other isotopic measurements. Again, it will be critical to increase our knowledge of the variance associated with isotopic ratios of the heavier elements in any given region of interest. In the short term, it would be most useful to test the utility of combining heavy and light isotope measurements to those species or populations that are reasonably tightly constrained geographically and that show the potential of occupying different geological substrates.

IV. SUMMARY

The intent of this volume was to provide the reader with a comprehensive background needed to understand the potential and the state-of-the-art in the application of stable isotope tools to the study of animal migration, and to encourage new research endeavors. Animal migration remains an exciting field that will provide many years of research for scientists in numerous disciplines. We have hopefully conveyed the idea that stable isotopes are not a "silver bullet" that will provide unambiguous insight into animal origins. The true potential of the isotope techniques will only be realized in cases where the researcher has been careful to first choose the species and migratory system that shows promise isotopically, and then considers the sources of variance in the model used to infer origins. The path ahead will involve far more emphasis on understanding the mechanisms that can influence isotopic variation spatially and within organisms of interest. It will also involve more careful consideration of how we statistically infer origins or establish the probability of assignment. Use of more refined isoscape models that involve several elements and the careful use of new remote sensing GIS layers will also be a fruitful area of research. Obviously, these areas of research and development are likely well beyond the scope of any single researcher or laboratory, and hence this field is sure to emerge as one of the best examples of multidisciplinary collaborative research.

V. REFERENCES

Bensch, S., S. Åkesson, and D. Irwin. 2002. The use of AFLP to find an informative SNP: Genetic differences across a migratory divide in willow warblers. *Molecular Ecology* **11**:2359–2366.

Evershed, R. P., I. D. Bull, L. T. Corr, Z. M. Crossman, B. E. van Dongen, C. J. Evans, S. Jim, H. R. Mottram, A. J. Mukherjee, and R. D. Pancost. 2007. Compound-specific stable isotope analysis in ecology and paleoecology. Pages 480–540 *in* R. Michener and K. Lajtha (Eds.) *Stable Isotopes in Ecology and Environmental Science*. Blackwell, Oxford.

Gannes, L. Z., D. M. O'Brien, and C. M. del Rio. 1997. Stable isotopes in animal ecology: Assumptions, caveats, and a call for more laboratory experiments. *Ecology* **78**:1271–1276.

Gleixner, G., and I. Mugler. 2007. Compound-specific hydrogen isotope ratios of biomarkers: Tracing climatic changes in the past.. Pages 249–266 *in* T. E. Dawson and R. T. W. Siegwolf (Eds.) *Stable isotopes as Indicators of Ecological Change*. Academic Press, London.

Guillot, G., F. Mortier, and A. Estoup. 2005. Geneland: A computer package for landscape genetics. *Molecular Ecology Notes* **5**:712–715.

Hobson, K. A., and E. Yohannes. 2007. Establishing elemental turnover in exercising birds using a wind tunnel: Implications for stable isotope tracking of migrants. *Canadian Journal of Zoology* **85**:703–708.

Irwin, D. E., S. Bensch, J. H. Irwin, and T. D. Price. 2005. Speciation by distance in a ring species. *Science* **307**:414–416.

Kelly, J. F., K. B. Ruegg, and T. B. Smith. 2005. Combining genetic and stable isotope markers to assign migrant songbirds to breeding origins. *Ecological Applications* **15**:1487–1494.

Koenig, W. D., D. Van Vuren, and P. N. Hooge. 1996. Detectability, philopatry, and the distribution of dispersal distances in vertebrates. *Trends in Ecology & Evolution* **12**:514–517.

Manni, F., E. Guerard, and E. Heye. 2004. Geographic patterns of (genetic, morphologic, linguistic) variation: How barriers can be detected by using Monmonier's algorithm. *Human Biology* **76**:173–190.

Pearson, S. F., D. J. C. H. Levey, and delRio, C. M. Greenberg. 2003. Effects of elemental composition on the incorporation of dietary nitrogen and carbon isotopic signatures in an omnivorous songbird. *Oecologia* **135**:516–523.

Popp, B. N., B. S. Graham, R. J. Olson, C. C. S. Hannides, M. J. Lott, G. A. Lopez-Ibarra, F. Galvanmagna, and B. Fry. 2007. Insight into the trophic ecology of Yellowfin Tuna, *Thunnus albacares*, from compound-specific nitrogen isotope analysis of proteinaceous amino acids. Pages 173–190 *in* T. E. Dawson and R. T. W. Siegwolf (Eds.) *Stable isotopes as Indicators of Ecological Change*. Academic Press, London.

Post, D. M., C. A. Layman, D. A. Arrington, G. Takimoto, J. Quattrochi, and C. G. Montaña. 2007. Getting to the fat of the matter: Models, methods and assumptions for dealing with lipids in stable isotope analyses. *Oecologia* **152**:179–189.

Salomonson, F. 1955. The evolutionary significance of bird migration. *Biologiske Meddelelser* **22**:1–62.

Webster, M. S., P. P. Marra, S. M. Haig, S. Bensch, and R. T. Holmes. 2002. Links between worlds: Unraveling migratory connectivity. *Trends in Ecology and Evolution* **17**:76–83.

Wikelski, M., R. W. Kays, J. Kasdin, K. Thorup, J. A. Smith, W. W. Cochran, and G. W. Swenson, Jr. 2007. Going wild—what a global small-animal tracking system could do for experimental biologists. *Journal of Experimental Biology* **210**:181–186.

Wunder, M. B. 2007. *Geographic Structure and Dynamics in Mountain Plover.* Ph.D. Dissertation. Colorado State University, Fort Collins, Colorado.

Wunder, M. B., and D. R. Norris. 2008. Improved estimates of certainty in stable isotope-based methods for tracking migratory animals. *Ecological Applications,* in press.

Index